FORWARD/COMMENTARY

This book, **THE EFFECTS OF RADAR ON THE HUMAN BODY** was published by the Department of Defense in 1962 and it describes how much damage Radio Frequency (RF) and Microwave (MW) radiation can do to the human body and indicates how much power is needed to incapacitate or even kill an individual. I believe this is a timely subject based on recent events such as the attacks on U.S. diplomats. There has been some speculation in the media that RF or microwaves may have caused the damage, however no one knows for sure.

What I do know is that RF and microwave radiation can be harmful so they should be studied and the literature needs to be updated. Personally, I don't use a wireless router at home because of the possibility of genetic damage or tumors.

The only copy I could find of this document was in pretty bad shape so I went ahead and scanned it and compared the scanned text with the original. Took several days to get it right. There were a few words that were totally obliterated, but that was the best copy available.

Why buy a book you can download for free?   We print this so you don't have to.

We at 4th Watch Publishing are former government employees, so we know how government employees actually use the standards. When a new standard is released, an engineer prints it out, punches holes and puts it in a 3-ring binder. While this is not a big deal for a 5 or 10-page document, many government documents are over 100 pages and printing a large document is a time- consuming effort. So, an engineer that's paid $75 an hour is spending hours simply printing out the tools needed to do the job. That's time that could be better spent doing engineering. We publish these documents so engineers can focus on what they were hired to do – engineering. It's much more cost-effective to just order the latest version from Amazon.com. SDVOSB and HUBZONE business.

If there is a standard you would like published, let us know. Our web site is www.usgovpub.com

Copyright © 2018 4th Watch Publishing Co. All Rights Reserved

Many of our titles are available as ePubs for Kindle, iPad, Nook, remarkable, BOOX, and Sony ereaders. Please visit our web site to see our recommendations.

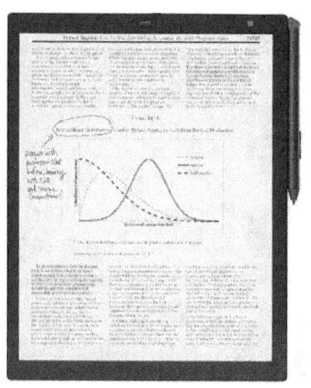  

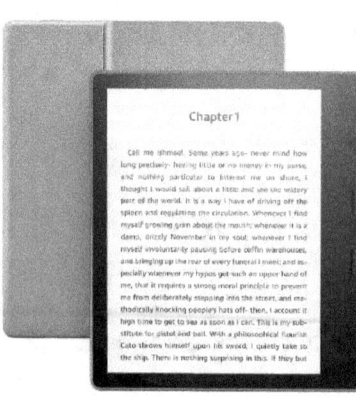  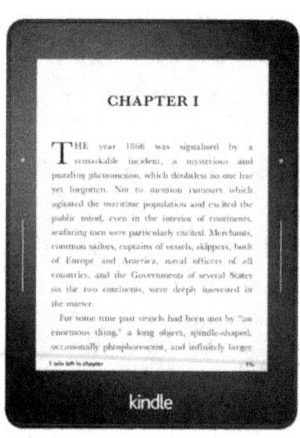

Why buy an eBook when you can access data on a website for free? HYPERLINKS

Yes, many books are available as a PDF, but not all PDFs are bookmarked? Do you really want to search a 6,500-page PDF document manually? Load our copy onto your Kindle, PC, iPad, Android Tablet, Nook, or iPhone (download the FREE kindle App from the APP Store) and you have an easily searchable copy. Most devices will allow you to easily navigate an ePub to any Chapter. Note that there is a distinction between a Table of Contents and "Page Navigation". Page Navigation refers to a different sort of Table of Contents. Not one appearing as a page in the book, but one that shows up on the device itself when the reader accesses the navigation feature. Readers can click on a navigation link to jump to a Chapter or Subchapter. Once there, most devices allow you to "pinch and zoom" in or out to easily read the text. (Unfortunately, downloading the free sample file at Amazon.com does not include this feature. You have to buy a copy to get that functionality, but as inexpensive as eBooks are, it's worth it.) Kindle allows you to do word search and Page Flip (temporary place holder takes you back when you want to go back and check something). Visit **www.usgovpub.com** to learn more.

21 March 1962 RM-TR-62-1

The Effects of Radar on The Human Body

by

John J. Turner
Resident AOMC ZEUS Project Engineer

U.S. ARMY ORDNANCE MISSILE COMMAND
Liaison Office, Bell Telephone Laboratories Whippany, New Jersey

Prepared By
Publications and Information Services Branch
Research and Development Directorate
Army Ordnance Missile Command

Reproduced

by the

**ARMED SERVICES TECHNICAL INFORMATION
AGENCY
ARLINGTON HALL STATION
ARLINGTON VIRGINIA**

NOTICE: When government or other drawings, specifications or other data are used for any purpose other than in connection with a definitely related government procurement operation, the U. S. Government thereby incurs no responsibility, nor any obligation whatsoever; and the fact that the Government may have formulated, furnished, or in any way supplied the said drawings, specifications, or other data is not to be regarded by implication or otherwise as in any manner licensing the holder or any other person or corporation, or conveying any rights or permission to manufacture. use or sell any patented invention that may, in any way be related thereto.

DESTRUCTION
This report shall be destroyed when no longer required,

REPRODUCTION
Reproduction of this document, in whole or in part, is prohibited except with permission of the issuing office; however, ASTIA is authorized to reproduce the document for United States Governmental purposes.

ADDITIONAL COPIES
Requests for additional copies of this document shall be addressed to Arlington Hall Station, Arlington, Virginia.

ABSTRACT

The probable effects of radio frequency radiation on the human body are discussed, and various published papers describing the effects of such radiation on biological subjects are briefly reviewed. The susceptibility of the head, the eye, and the testis to RF radiation is given separate coverage. Ionizing radiation produced by RF generating equipment is also discussed. The document draws no conclusions and makes no recommendations.

ACKNOWLEDGMENT

Appreciation is extended to those investigators whose work is quoted herein and apologies to those whose work should have been quoted or referred to but was not, in the interest of brevity. Grateful acknowledgment is made to Bell Telephone Laboratories for the use of their Technical Library and facilities, to Mr. W. W. Mumford of the Laboratories for his helpful criticism and suggestions, to Lt Col R. C. Miles and Lt Col L. G. Jones for their encouragement and guidance and to Mrs. J. G. Shelby, Mrs. P. H. Herold, Miss M. A. Vreeland and Miss P. A. La Bruto for their assistance in preparation of the manuscript. Appreciation for final editing is due Mr. L. S. Stauffer of the Army Ordnance Missile Command.

PREFACE

Increasing use of radar and other microwave generating equipment by the military services and the ever-increasing power of such equipment has made evident the need for a single source of information on the effects of radio frequency radiation on the human body.

Although studies of the effects of such radiation on biological specimens have been described in reports since 1927, the literature has become so extensive and involves so many different scientific fields that specific information concerning the possible hazards to personnel operating high-powered microwave equipment is not readily available to supervisory personnel charged with the safe operation of such equipment.

This document is a report of some of the significant investigative work that has been performed in this field. An effort has been made to make it brief and easily understood. No conclusions have been drawn and no recommendations have been made. It should be borne in mind that safety provisions covering U. S. Army personnel in connection with microwave radiation hazards are outlined in AR 40-583 entitled, "Hazards to Health from Microwave Energy," and that establishment of such safety provisions is properly within the province of the U. S. Army Surgeon General.

TABLE OF CONTENTS

Chapter		Page
1	History	1
2	Calculation and Measurement of Microwave Energy	7
	The Microwave Region of the Electromagnetic Spectrum	7
	Radiation from Radars	8
	Zones within a Radar Field	8
	The "Near Field" or Fresnel Region	8
	Irradiation of Specimens in the "Near Field"	9
	Cages and Containers	10
	The Intermediate Field	11
	The "Far Field," or Fraunhofer Region	11
3	Microwave Measuring Instruments	15
	Field Strength and Power - Level Meters	15
	Pocket - Type Dosimeter	17
	The Ion Orb Indicator	18
	Other Problems of Measuring Human Irradiation	18
4	Whole Body Irradiation	20
	The Relative Absorption Cross Section	20
	Dielectric Properties of Body Components	22
5	Irradiation of the Head	27
	Auditory Response	27

TABLE OF CONTENTS- Continued

Chapter		Page
	Pulsed Energy Sleep	28
	Irradiation of the Head of a Monkey	30
	Additional Experiments with Monkeys	31
	Irradiation of the Heads of Dogs	32
	Irradiation of the Heads of Rats	35
6	Irradiation of the Testis	37
	Effects of Microwave Irradiation at 2880 mc (10.4 cm)	37
	Effects of Microwave Irradiation at 24,000 mc (1.25 cm)	38
	Comparison of the Effects of Microwave Radiation at 2450 mc and the Effects of Infrared Radiation	40
	Comparison of the Effects of Microwave Radiation at 24,000 mc and the Effects of Infrared Radiation	42
7	Effects of Microwave Irradiation of the Testes on the Endocrine System	43
	Location of Damage in the Endocrine Chain	45
	Microwave Exposure Versus Infrared Exposure	46
8	Irradiation of the Eye	48
	Time and Power Thresholds for Induction of Lens Opacities by Continuous Wave Radiation at 2450 mc	49
	Cumulative Effects of Repeated Subthreshold Exposure Periods	49
	Effects of Pulsed Microwave Radiation	53
	Non-Complementary Effects of Microwave Radiation and X-ray	54

TABLE OF CONTENTS - Concluded

Chapter

	Comparison of Opacities Resulting from Microwave and Infrared Radiation	55
	Study of Possible Interface Effect at Lens - Vitreous Body Boundary	55
	Effect of Distance of Eye from Microwave Source	56
	Changes in the Ascorbic Acid Content in Lenses of Rabbit Eyes Exposed to Microwave Radiation	57
9	The Effect of Microwaves on Unicellular Organisms	59
	Review of Past Experiences	59
	Experimental Cell Research Using Pulsed Electromagnetic Fields	61
10	Generation and Detection of Ionizing Radiation Produced by Microwave Equipment	67
	Generation of X - radiation	69
	Detection of X - radiation	71

Chapter 1

HISTORY

It seems to be an established fact that, in both Egypt and India, people knew how to make use of electromagnetic radiation some 3,000 years ago.[1] Because they had no radiation-producing equipment, they used sunlight. The experimental material was the human skin, and the instigator of these experiments was the science of medicine. Skin conditions believed to be vitiligo were customarily treated by painting the skin with a concoction and exposing it to sunlight. This concoction has been found to contain furocoumarins which are known to convey the photodynamic action in bacteria. Light alone has no effect, and the coumarins alone have no effect on the killing of bacteria, but the combination of the two agents is most effective. Energy absorbed by one molecule is conveyed to a biologically important molecule, which is changed, and thus lethal damage is inflicted on the cell.[2] Vitiligo is-a skin disease manifested by smooth, milk-white spots on various parts of the body. Coumarin is a white, crystalline compound ($C_9H_6O_2$) of vanilla-like odor, found especially in the tonka bean and used in flavoring and perfumery.

In 1791 Galvani published some observations on frogs' legs. He noticed that in contact with copper and iron they twitched vigorously, and he was unable to fathom a reason for this effect. One hundred and seventy years have passed since Galvani noted the phenomenon and a satisfactory explanation cannot yet be given. Much progress has been made but much remains obscure, particularly in the areas of biophysics and physical chemistry.[3]

It was pointed out in 1936 by Professor Szymanowski of the Institute for Physiological Research, Moscow, U.S.S.R. that in 1896 D'Arsonval and Charrin reported attenuation of diphtheria toxin by radiation of frequency 2×10^5 cycles per second without significant temperature elevation.[4]

In 1908 von Zeyneck grasped the possibilities inherent in heating body tissues by the conduction of a high - frequency current through them, a therapeutic method for which he coined the name "diathermy."[5] His technique was to apply electrodes to the skin of the body. This was a rather ineffective method since the various tissues have markedly differing resistances and consequently the heating was not uniform. As the frequencies available increased with technological advances, it became apparent that generation of heat in tissues depends on frequency.

Beginning in 1928, Schliephake[5] employed higher frequencies in the 30 mc region for therapeutic experiments. At these higher frequencies, the predominant effect was one of induction rather than conduction, and a much more uniform depth penetration could be achieved. It became possible to heat the deeper layers without overheating the skin and the subcutaneous tissues.

The history of the use of radio waves reveals many attempts to employ them experimentally for the generation and investigation of poorly understood phenomena. The methods employed by some of the early research workers in this field were of somewhat doubtful validity, as were some of the conclusions they drew from their results. Yet these early investigators deserve mention and credit for their work; some of it is being repeated with modern methods, under carefully controlled conditions.

For example, the Italian physician Cazzamalli attempted in the 1920's to elicit heterodyned radiation from human brains by exposing them in vivo to the field of a relatively powerful radio transmitter.[6] He claimed to have observed and recorded a variation in such radiation when his subjects were emotionally aroused or engaged in creative pursuits. Among his results was the finding that hallucinations could be induced in highly suggestive individuals by radiation.

In 1936 Krasny-Ergen[15] advanced theories on the behavior of coloids in alternating electric fields and the effect of the inductive forces between the dipoles and multipoles which are induced by such fields in dispersed particles. The particles could be small living or dead organisms, or liquid or gaseous particles dispersed in liquids or gases. He also proposed a theory for "pearl-chain" formation of dispersed particles and pointed out the importance of this phenomenon in the field of biology, particularly with respect to the reactions between antigens and antibodies and the effects on cell division and fertilization. He also examined the effects of these fields on the viscosity of colloids and found them to be dependent on the direction of the field and on whether the field was a rotating field, a field of constant direction, or a combination of both. He also showed that rotating fields occur spontaneously at certain frequencies in the short and ultra-short wave range and that these fields could be biologically significant.

The results reported by the French physician Lakhowski are very questionable. He believed that multifrequency radiation introduced in the vicinity of his subjects was responsible for their recovery from malignant growths and concluded that this type of radiation contained the key to the secret of life.

Considerably more plausible are the results produced with microwave radiation by a Dutch physician, Van Everdingen, a tireless, careful, persistent and imaginative research worker. He reported[8] the reduction of certain types of growths by carefully planned procedures that included injecting irradiated substances. The results included control of the appearance of growths in healthy animals exposed to time-tested, coal-tar sti   s and the induction of growths by the reversal of certain steps   procedure. Van Everdingen observed that microwaves affected the heart action of chicken embryos and that this effect did not occur until glycogen first appeared in the affected hearts. He concluded that some action on the glycogen was responsible for the result. Further experimentation revealed that the microwaves produced a measurable change in the plane of optical polarization in the glycogen if the concentration and viscosity of the substance were precisely controlled. The amount by which the polarization plane was rotated proved to be an accurate indicator of dosage of a given frequency. Van Everdingen concluded that it was this kind of "unnatural" mechanism that controlled tumor growth. Extracts from the livers of healthy mice, when irradiated, would affect the resistance to tumor growth of mice injected with the altered substance. It is understood that the U. S. Army Medical Research Laboratory at Fort Knox, Kentucky is presently engaged in extending some of Van Everdingen's early experiments.

Numerous other experiments have been performed to study the nonthermal effect of radio waves on biological specimens. As an example, Nyrop reported in 1946[9] specific effects on bacteria, viruses, and tissue cultures resulting from experiments in which heating was carefully excluded by applying the radiation in short pulses, with intervals long enough (and increasing as the experiment progressed) to prevent the temperature from increasing. Although experiments of this sort have elicited considerable skepticism in the past, more recent results suggest that a physical basis may indeed exist for some of the earlier observations.

Another aspect that has attracted attention in scientific circles in this country and in the Soviet Union is the possible neurological effect of electromagnetic radiation. One Soviet article[10] cites many results that cannot be dismissed as thermal effects. The Soviet scientist, Gordon, has reported that after irradiation with microwaves at low power density (5 to 10 mw/cm^2), the delay between stimulus and conditioned reflex in dogs increased, and the employment of a larger stimulus became necessary. In addition, histological examinations of brain tissues revealed physiological and chemical changes such as globular concentrations of acetylcholine along nerve fibers.[11]

Military radars are generally designed to operate in what is known as the "microwave" region of the radio-frequency spectrum. This range includes frequencies from approximately 1000 mc to 30,000 mc or higher. Expressed in wavelengths, this region is from 30 cm to 1 cm.

Microwave energy is generally produced by means of specially constructed electronic tubes such as the magnetron and the klystron. The magnetron was invented by Albert W. Hull of the General Electric Research Laboratories in 1920. The klystron was developed by the Varian brothers and W. W. Hausen working under D. L. Webster of Stanford University in 1938.

Physicians at the Mayo Clinic became interested in microwaves in 1937. At that time, Krusen, Hemingway and Stenstrom, in the department of physics at the University of Minnesota in Minneapolis felt that such radiation might be particularly valuable in medicine if it could be obtained in sufficient intensity. Investigation disclosed, however, that power of only 2 or 3 watts could be produced by magnetron tubes then available. This power was far below that required for therapeutic purposes.

By July 1938, the magnetron was capable of producing at least 100 watts at any wavelength down to 20 cm.

In March 1939, D. L. Webster, one of the developers of the klystron tube advised that the klystron could produce several hundred watts at wavelengths between 10 cm and 40 cm.

Just when tubes of sufficient power had finally been found, all such tubes suddenly became mysteriously unavailable. It was not until the secret of radar was finally revealed that it became evident such tubes had been designated for military use only.

Although unknown to research workers in the medical field, about that time a research group at the University of Birmingham in England developed the multicavity, air-cooled magnetron tube which was of tremendous importance in perfecting radar, which, it has been said, "won the battle of Britain."

In September 1940, a British technical mission headed by Sir Henry Tizard brought the multicavity magnetron to the United States and in a short time American manufacturers were producing the tube for use in microwave radar. The output of these tubes could be as high as one million watts and the microwaves they produced had optical properties so that they could be reflected, refracted, or diffracted.

In 1946 one of these microwave generators became available at the Mayo Clinic and studies on living animals were begun by Drs. U. Leden, J. Herrick, and K. Wakim. Other workers included Drs. R. E. Worden, J. W. Gersten, J. W. Rae, Jr., J. P. Engle, L. Daily, Jr., and R. Q. Royer. The first paper on the effects of microwave radiation was published in the Proceedings of the Mayo Clinic, 28 May 1947,[12]

Considerable impetus was given to the study of the biological effects of microwave radiation, and in particular to the possible injurious effects to the human body by an article by Dr. J. T. McLaughlin published in the May 1957 issue of the magazine, California Medicine, entitled, "Tissue Destruction and Death from Microwave Radiation (Radar)." The article was a case report on the death of a man who stood in the direct beam of a radar transmitter. In a few seconds the man had a sensation of heat; the heat became intolerable in less than a minute and he moved away from the antenna. Within 30 minutes he had acute abdominal pain and vomiting. Several operations were performed but he died within ten days from inflammation of the intestines attributed to destructive heat generated by the radar beam. It has not been possible to determine the power density involved in this particular case of exposure but it is believed that the 6249 magnetron and the APG-37 fire control radar were involved. This equipment is capable of delivering 300 watts average power. The antenna size and configuration could not be determined.

In July 1957, in connection with microwave radiation hazards, the Chief of the Research and Development Division, Ordnance Missile Laboratory, Redstone Arsenal, Huntsville, Alabama, stated, "A definite problem area is present with high powered radars and is increasing with increased power of the radars." Other statements made at about that time follow. Sylvania Electric Products Company; "The amount of solid information available today as to the effects of high powered radars on men and explosives is distressingly small." Bell Telephone Laboratories stated, "A real problem exists which may become acute, the safe limit of power density should be determined."

Originally, the military responsibility for the study of the biological effects of microwave radiation was assigned to the Navy and the studies were carried out at the Naval Medical Research Institute. At the request of the Navy, the responsibility was transferred to the Air Force, and the radiation biology program was conducted at the School of Aviation Medicine, Randolph Field, Texas. By 1956 the military services recognized the seriousness of the problem in connection with radars and a tri-service study program was instituted. Coordinating responsibility was delegated to the Air Force, and technical direction of the program was assigned to the Rome Air Development Center, Rome, New York. Since that time, five tri-service conferences have been held on the subject in July 1957, July 1958, July 1959, July 1960, and July 1961.

Chapter 2

CALCULATION AND MEASUREMENT OF MICROWAVE ENERGY[13,14]

Before entering into a discussion of the calculation and measurement of microwave energy, it would seem appropriate to examine the trend of development in magnetrons and klystrons which produce this energy. W. W. Mumford of Bell Telephone Laboratories has pointed out[13] the importance of being aware now of the hazards of the future. In 1940, 10 watts of average power was available. By 1945, magnetron improvements made possible the production of 1.1 kw of average power. By 1957 the klystron was capable of delivering 8.0 kw of average power. It is estimated that by 1965 at least 1,000 kw of average power will be available. The trend from 10 watts in 1940 to 1,000 kw in 1965 represents a rising capability of 50 db in less than 30 years, or a rate of over 15 db per decade. It is therefore apparent that we must closely examine the problem of microwave radiation hazards today lest we be faced with an intolerable situation in the future.

THE MICROWAVE REGION OF THE ELECTROMAGNETIC SPECTRUM

The microwave region is considered to extend from the highest radio frequencies down to the ultra-high frequency band between 300 and 3,000 megacycles per second but radiation hazards may exist at any radio frequency capable of being absorbed by the body.

The radio frequencies include eight regions corresponding to the eight decades of wavelength they occupy. The eight bands are as follows:

(a) Very Low Frequency (VLF) - 10^7 to 10^6 cm

(b) Low Frequencies (LF) - 10^6 to 10^5 cm

(c) Medium Frequencies (MF) - 10^5 to 10^4 cm

(d) High Frequencies (HF) - 10^4 to 10^3 cm

(e) Very High Frequencies (VHF) - 10^3 to 10^2 cm

(f) Ultra—High Frequencies (UHF) - 10^2 to 10 cm

(g) Super-High Frequencies (SHF) - 10 cm to 1 cm

(h) Extra High Frequencies (EHF) - 1 cm to 10^{-1} cm

RADIATION FROM RADAR

Both radar and communication systems may produce hazardous electromagnetic power densities (average watts per sq cm). Radar systems are generally characterized by pulsed operation and scanning antenna beams, while communication systems are generally continuous wave in nature and usually have fixed antenna beams.

Acquisition and search radars are normally used only when the antenna is scanning and the average power absorbed by an object at a fixed point is therefore reduced by the fact that the direct beam is pointed in that direction only a fraction of the time it takes for the antenna to complete one revolution. Some of these radars produce such a strong field that if they were not rotating, the power density might be hazardous to a distance of 500 ft or more. In such case, interlocks may be used to insure that the transmitter is not operating while the antenna is stationary.

Tracking radars do not scan but point toward the target or the missile. They are therefore potentially more hazardous than the acquisition or search radars even though their power may be less. In many cases it may be necessary to provide interlocks to prevent radiation from a tracking radar in certain critical directions.

ZONES WITHIN A RADAR FIELD

The field in front of the usual parabolic radar antenna is generally divided into the three following regions:
(a) The "near field," or Fresnel region, where the radiation is substantially confined within a cylindrical pattern.
(b) The "intermediate field" - a transition zone in which the power density decreases with increasing distance but not in accordance with the inverse square law.
(c) The "far field," or Fraunhaufer region, beyond the Fresnel and intermediate regions in free space, where the radiation is essentially confined to a conical pattern and the power density along the beam axis falls off inversely with the square of the distance.

THE "NEAR FIELD" OR FRESNEL REGION

Since most investigators do not have sufficient power available to conduct experiments with the biological specimen exposed in the far field, it has been necessary in most cases to bring the specimen within the near field.

J. H. Vogelman of the Capehart Corporation has observed[16] that two phenomena are present in the near field which contribute to errors in the determination of the actual field density to which a biological specimen is exposed and, in turn, introduce a questionable factor in the quantitative values for observed effects. The first effect is the cyclic variation in field density in the near field as one proceeds from the antenna aperture outward to the end of the near field region. The exact position of the specimen with respect to the antenna aperture will determine the ambient field density. Because the introduction of the specimen into the field moves the cyclic variation, it is almost impossible to predict the actual ambient field density within the near field for any but the simplest states suitable for exact or good approximate computation. At the same time, measurements of the field density in the near field may not suffice since the field variations are displaced to a different degree by the measuring instrument and the specimen. Where the measuring instrument may indicate a peak, the introduction of the specimen at the same point may result in a minimum of ambient field density.

The second phenomenon which results from the introduction of a biological specimen into the near field of an antenna is an interaction between the specimen and the antenna. This results in an impedance mismatch as seen by the generator of the microwave power. This mismatch may result in a marked change in the power generated as well as in a change in oscillator frequency. These effects depend on the degree of sensitivity of the generator to standing waves and the proximity of the biological specimen to the antenna aperture. Accordingly, near field measurements of biological effects lack quantitative accuracy since the prediction of the ambient field density is both difficult and inaccurate.

IRRADIATION OF SPECIMENS IN THE "NEAR FIELD"

Vogelman has recommended[16] two approaches to obtain more accurate measurements of power density in the near field region. The first approach, intended for whole-body radiation of biological specimens would use a metallic chamber of good conducting material such as copper screening of 20 X 20 mesh to house the specimen to be radiated. The chamber would be connected directly to the transmission line coupled to the transmitter signal source. A bidirectional coupler would be incorporated in the transmission line to provide indication of the incident power as well as the reflected power. With the chamber empty, the highly conductive walls would reflect the radio frequency energy with the result that the reflected power would be essentially identical with the incident power. Where required, a Faraday rotation isolator could be inserted between the signal source and the bidirectional coupler to ensure that the reflected energy would not cause breakdown in the signal source. Such a device should be set up to provide a load for the reflected power away from the signal transmitter tube itself. With the specimen inserted into the chamber, the difference between the incident power and the relative reflected power would be the energy absorbed by the specimen. If the incident and reflected power outputs from the directional coupler were recorded, the animal would be free to wander about the chamber, and a record of instantaneous exposure would still be available. Such a chamber should be nonresonant at the frequency of operation.

For the specific radiation of a single appendage or area of a biological specimen, Vogelman[16] recommends the following arrangement: The output of the waveguide structure is terminated either in an expanded section or in an extremely thin-walled iris with an opening the size of the area to be exposed to microwave energy. Spring fingers are used around the waveguide structure to ensure good coupling to the specimen and at the same time minimize leakage radiation. A set of tuners are included in the waveguide structure. The specimen is inserted into the spring fingers so that the enlarged section of the waveguide or the iris is in contact with the appendage or portion of the body to be irradiated. The tuners are adjusted so that the reflected power is reduced as close to zero as possible. The difference between the incident power and the reflected power is the energy coupled to the biological specimen.

CAGES AND CONTAINERS

Vogelman[16] has outlined basic rules which must be observed in the use of cages, containers, or other devices for securing biological specimens. Such should be fabricated of dielectric materials having the lowest possible loss. He suggests Polystyron, polyvinal chloride, or Teflon as suitable materials. The spacing between components of the dielectric (the open spaces) must be at least one wavelength in the direction perpendicular to the polarization of the antenna or waveguide. The use of water coolant must be so confined as to provide an unobstructed path at least one wavelength in diameter between the waveguide the specimen. Otherwise, the water will completely shield the specimen from the microwave radiation. In addition, when absorbing material is used to shield portions of the specimen, the absorbent must be between the exposed area and the source of microwave energy. If the specimen is extended through the absorbent it will be exposed to direct thermal heating from the absorbent as well as microwave heating, resulting in unreliable data.

THE INTERMEDIATE FIELD

Between the end of the near field and the beginning of the far field of a radar antenna there lies a transition zone in which the power density decreases with increasing distance but not in accordance with the inverse square law. This zone has been called the quasi-Fresnel, or "crossover" region. H. S. Overman of the U. S. Naval Weapons Laboratory[17] has proposed the following empirical equation for obtaining an approximation of the power density in the intermediate field:

$$p = 0.87 \, (w/\lambda r)$$

where

p = power density

r = distance from antenna

λ = wavelength of radiation

w = average power delivered to antenna

THE "FAR FIELD," OR FRAUNHOFER REGION[17]

The power now in a radar beam at a considerable distance away from the antenna, in what is generally called the "far field" may be thought of as confined within a cone which has its apex at the antenna. The apex angle of the cone is the "beam width." The cross-sectional area of the beam varies with the square of the distance from the antenna; hence the power density, which is the power per unit area, will be proportional to the reciprocal of the square of the distance, i.e., it will conform to the inverse square law. In practice, however, the power radiated by an antenna is not all confined within the conical beam. Some is radiated just outside the nominal limits of the beam and some in side lobes. In addition, the power density is about twice as great on the beam axis as at the edges. The larger the antenna, the higher the "concentration" of power. This "concentrating" action of an antenna is called the antenna "gain." A large antenna with a narrow beam thus has a large gain. The "gain" is a pure number which can always be furnished for each antenna.

To compute the power density at the beam center in the Fraunhofer region, Mumford[13] suggests the use of the following formulae:

$$w = \frac{GP}{4\pi r^2} = \frac{AP}{\lambda^2 r^2}$$

where

w = power density

G = antenna gain

P = average power output (not peak power)

D = diameter of antenna

A = area of antenna

Using a power density equal to the potentially hazardous level of ten milliwatts per square centimeter as specified in AR 40-583, Mumford[13] has collected data on the distance to the boundary of the potentially hazardous zone for several radars. These distances are listed in Table I.

TABLE 1

DISTANCE IN FEET FROM RADAR ANTENNA TO BOUNDARY OF POTENTIALLY HAZARDOUS ZONE FOR SOME COMMON RADARS (ARRANGED IN DESCENDING ORDER OF DISTANCES)

Radar type	Distance for 0.01 watt/cm², ft
AN/FPS-16	
Sig C Mod.	1020
Standard Mod.	590
AN/FPS-6	560
HERCULES Improved Acq.	
HIPAR (Fixed)	550 note 1, 2
AN/MPS-23	530
AN/MPS-14	472
HERCULES Imp TTR	400
AN/TPQ-5	350
AN/FPS-20	338
AN/MPQ-21 (10')	300
HERCULES MTR (AJAX)	270
AJAX Acq. (Fixed)	260 note 1
AN/CPS-9	260

TABLE 1 – Continued

Radar type	Distance for 0.01 watt/cm^2, ft
AN/MPQ-21 (7')	210
AN/MPS-4	205
AN/FPS-8 (40' x 14')	205
AJAX MTR	205
AN/CPS-6B	200
AN/FPS-10	200
AN/MPS-22	185
AN/FPS-18	178
AN/MPS-12	175
AN/MPQ-18	175
AN/FPS-3	172
AN/MPS-7	172
AN/MPQ-21	165
AN/TPS-1G (40' x 11')	150
AN/FPS-36	150
AJAX TTR	132
HERCULES Improved Acq. (Fixed)	130 note 1, 2
HERCULES Acq. (Fixed)	130 note 1, 2
AN/FPS-14	109
AN/FPS-4 (narrow pulse)	106
AN/MPS-8 (narrow pulse)	106
AN/TPS-10D (narrow pulse)	106
AN/MPS-10 (C)	105
AN/FPS-8	101
AN/MPS-11	101
SCr 584	70
AN/MPQ-10 (S)	50
AN/TPS-1-D	50
AN/FPS-25	40
AN/FPS-31	27.5
HERCULES Improved Acq. HIPAR (Rot.)	25
AJAX Acq. (Rot.)	8
AN/PPS-4	2.5

TABLE 1 – Concluded

NOTE 1: Not normally used with fixed antenna.
NOTE 2: Interlocks provide assurance that transmitter is idle unless antenna is rotating. Calculated distances in Table I are based on the following assumptions:

(a) Free space transmission.

(b) No ground reflections. These could double the distance shown.

(c) Calculations apply to the axis of the beam, i.e. where the power density is greatest.

(d) The beam is considered to be fixed in space, i.e., not scanning.

Chapter 3

MICROWAVE MEASURING INSTRUMENTS

The subject of microwave measuring instruments is extensive and complex, and for that reason, this chapter will be essentially limited to those instruments used in "dosimetry." By "dosimetry" is meant the measurement of radiation actually received and active in the subject's body.

The radiation received on the surface of the body can be measured for certain frequency ranges with reasonable accuracy. For this, laboratory-type instruments must be used which are unsuitable for wear by a person who must work around radiation-producing equipment. The matter becomes even more acute at microwave frequencies, such as those to which radar technicians are exposed; where even the accurate measurement of energy arriving on the surface is a considerable problem. No accurate, portable (pocket-type) dosimeter which has sufficient versatility, that covers enough frequencies, and is usable on all people small or large, fat or thin is presently available. Many attempts have been made to build such instruments, but in the final tests, they always fall short of the requirements.

Basically, the instruments currently available for dosimetry can be divided into two types.[18] One is the field-strength meter, familiar to every TV technician and radio amateur. The second basic type is the power-level meter. A third type of radiation meter, the echo chamber, need not be considered here because it is useful only at relatively high power levels. There are also some chemical substances which will show thermal effects from radiation and will change color. However, these—substances are very unstable and rather sensitive to humidity and must be considered impractical at the present time.

In addition to available types, some recently developed measuring instruments are also described here, together with a summary of additional problems of measuring human irradiation.

FIELD STRENGTH AND POWER-LEVEL METERS[18]

A field strength meter is principally a sensitive receiver, usually a superheterodyne, with a meter which indicates the strength of the signal received. Sometimes field-strength meters are very simple affairs, with a tuned circuit and diode rectifier before the meter. These can be considered "receiver" types.

A power-level meter is usually a sensitive bridge which uses a <u>transducer</u> to convert electromagnetic energy into another form of energy and to produce a change due to thermal effects. A transducer which translates thermal effects is called a <u>bolometer</u>. Heat due to energy absorbed from the RF field results in a change in the electrical resistance of a bolometer, thereby permitting measurement of the RF field density. The bolometer used with a power-level meter may be either a <u>thermistor</u>, which decreases in resistance as the temperature rises, or a <u>barretter</u>, which increases its resistance as the temperature rises. An electric current will vary inversely as the resistance, and will be registered on the meter, so measuring the RF energy which produced the heat. A <u>thermistor</u> is a bead of metallic oxide (which has a negative resistance coefficient) with two wires inserted in it. The bead is subjected to high temperatures, sintered, and then assumes semi-conductor properties. It is more sensitive than other bolometers.

A <u>barretter</u> is a very thin piece of platinum wire in a suitable mounting. The wire is far too thin to be seen with the naked eye. The barretter is constructed of a thin platinum wire surrounded by a sheath of silver. This combination is then rolled out to a very small diameter, often less than a thousandth of an inch. This thin, combination wire is then mounted in a suitable holder, after which the silver sheath is carefully etched away, leaving the microscopically thin platinum wire. The barretter responds much more rapidly than a thermistor. It can also be burned out readily by overloads.

Certain crystal diodes (1N21, 1N23, etc.) are available for transducer service. The crystal transducer, when subjected to electromagnetic radiation, generates a small voltage analogous to the generation of voltage in a silicon solar battery by light.

All of these devices are sensitive to temperature. In instruments used to measure an RF field, any temperature variation in the environment or in the equipment involved in the test must be taken into consideration.

A typical, commercial, power-level meter is that manufactured by Empire Devices Products Corporation. The accessories for this instrument include a connecting cable, a dipole antenna, and microwave horns used to pick up the RF energy. The antennas will cover frequency range from 200 to 10,000 mc. The dipole is adjustable and can be tuned from 200 to 800 mc. Each horn will respond to a broad range of frequencies. The most sensitive range of this power-level meter is 2 milliwatts per square centimeter, one-fifth of the specified hazardous dosage for human subjects.

The Empire meter is basically a Wheatstone bridge using thermistors as bolometers. The instrument is simple and reliable but is somewhat limited in sensitivity.

A more elaborate power-level meter designed for microwave measurements is the Hewlett Packard, Model 430C microwave power meter. Basically, this instrument consists of a bridge, an oscillator, a v.t.v.m., and a regulated power supply. The bridge is an a-c type with power supplied by the variable-output oscillator. The amount of power supplied is inversely proportional to the unbalance in the bridge - - the greater the unbalance, the lower the power. This instrument can use either a thermistor or a barretter as a bolometer.

Two other devices designed specifically for microwave power density measurement have recently appeared on the market and are favorably regarded by W. W. Mumford of Bell Telephone Laboratories. They are the RAMCOR "Densiometer," Model 1200, and the Sperry Radiation Monitor; Model B86B l.

The "Densiometer", Model 1200 is manufactured by Radar Measurements Corporation, Hicksville, New York. It covers 5 bands with 4 different antennas: VHF (200-225 mc), UHF (400-450 mc), S Band (2600-3300 mc), C Band (5000-5900 mc), and X Band (8500-10,000 mc).
The power density for mid-scale reading is 10 mw/cm^2.
It weighs about 2 pounds.

The Radiation Monitor, Model B86Bl is produced by Sperry Microwave Electronics Company Clearwater, Florida and covers from 200 to 10,000 mc. Interesting and desirable features of this instrument are that it receives all polarizations simultaneously and requires but one antenna. The power density for mid-scale reading is 10 mw/cm2. It weighs 2 pounds.

POCKET-TYPE DOSIMETER[18]

In contrast to the laboratory-type instruments described above, a pocket-type dosimeter has been developed by Jaski,[18] in accordance with data supplied by Dr. A. W. Richardson of St. Louis University. This instrument consists of a simple, resistive-type antenna, a diode detector, an a-c amplifier, a second detector, and a meter. The amplifier can only amplify a-c; the instrument is therefore limited to the measurement of <u>modulated</u> microwaves whose modulation frequency lies within the response capabilities of the amplifier. Unmodulated RF energy will not be indicated on the meter.

This instrument has some obvious disadvantages due to the simplicity of its circuitry. It could possibly show the same radiation level for what are actually different levels modulated by a different wave shape because of the limited response of its amplifier. Continuous wave energy will not be registered at all. In addition, the instrument is sensitive to hand capacity and will pick up "noise" - - stray radiation from such sources as fluorescent lamps, etc.

Unsatisfactory as this instrument is, it is a step in the right direction. The military services and industry are actively seeking a pocket-type dosimeter which will do the job of warning personnel of the presence of excessive radiation. Such an instrument must meet certain criteria; it must be compact, accurate, sensitive, frequency insensitive within a broad range, stable, and preferably inexpensive. This obviously presents a problem of considerable magnitude.

Compounding the problem is the conclusion arrived at by Professor H. P. Schwan of the University of Pennsylvania, who has stated,[19] "The variability of impedance match of the human body surface to air makes it obvious that any sensing device must be used at a 'sufficient' distance from the human body. How large the distance should be is impossible to state at present in the absence of pertinent data, but it most certainly is sufficiently large to prohibit the device being carried on the body surface."

THE ION ORB INDICATOR

H. R. Meahl of the General Electric Company recently announced[20] the development of an ion orb, omnidirectional, fixed level, visual indicator of radio frequency field strength. The indicator responds to peak power and consists of a glass sphere filled with a combination of helium and neon gas. An ion orb 4 inches in diameter will glow red and orange in a field of 10 to 12 volts rms per cm. Orbs 1 inch in diameter will glow at 30 to 34 volts rms per cm. Meahl feels that these indicators will be useful in RF field strength hazard alarm systems over the range of 50 to 500 mc and probably to 3,000 mc. They weigh only 5 ounces, are omnidirectional, and had stable characteristics over a period of one year under field conditions in the tropics. They also operated satisfactorily after overloads which heated the glass to approximately 50° C.

OTHER PROBLEMS OF MEASURING HUMAN IRRADIATION[20]

Even after the radiation to which a subject has been exposed has been measured, the problem is only half solved. In the first place, not every subject is affected by radiation to the same extent. Frequency of the radiation is also an important factor. At some frequencies, people with thick layers of subcutaneous fat absorb radiation readily, while at other frequencies the fat acts as a virtual insulator. Ultra-high frequencies through the body and can heat internal body tissues but layers of fat will obstruct the radiation.

With microwaves, the problem becomes more difficult. Some of the microwave energy is reflected at the body surface - - the amount of reflection depending on the body surfaces, the polarization of the energy, and the intervening medium between the generator and the subject. Microwave energy can be readily absorbed by the skin and fatty tissue directly beneath the skin and will not heat the body tissues underneath directly but as a result of stepped-up circulation of the blood.

Thus, although the dielectric properties of body tissues are well known and have been measured with some degree of accuracy, we cannot predict the effects of radiation on an individual without knowing a great deal about his physical make-up. Moreover, in considering nonthermal effects, the problem becomes even more complex. Van Everdingen has sought to eliminate this problem by inserting, subcutaneously, small capsules of glycogen of precise concentration and viscosity.

The glycogen shows a rotation of its optical plane of polarization, which can be measured with a polarimeter. This procedure is obviously unsuited for use with human subjects, although it did provide accurate dosage measurement for test animals.

Chapter 4

WHOLE BODY IRRADIATION

THE RELATIVE ABSORPTION SECTION

In considering exposure of the body to radio frequency radiation it is of importance to determine the relative absorption cross section of the entire body for this type of radiation. The foremost investigators in this particular field are Schwan, Satati, and Anne of the University of Pennsylvania. Their latest report of investigation[21] describes their experiments using hollow, plastic dolls to simulate the shape to predict absorption. They define the relative absorption cross section as the ratio of the power absorbed from the incident field to the power incident on the object's geometric cross section prior to its insertion into the field.

All body tissues within the following range of electrical values throughout the total microwave frequency range:

   Relative Dielectric Constant = 5 to 70
   Resistance = 10 to 10,000 ohms/cm
   Relative Permiability = 1

Electrolyte mixtures of Dioxane potassium chloride in water having suitable combinations of properties in the above range were developed and used to fill the plastic dolls for the experimental work. The results obtained are listed in Table II.

The relative absorption cross section is a function of the conductivity of the exposed material as well as the size of the object relative to the wavelength of the incident field.

For objects much larger than the wavelength, the relative absorption cross section remains below unity and decreases with increasing size of the object. For objects small compared with the wavelength, the relative absorption cross section can become greater than unity. For this case, the object absorbs more power than is incident on its geometric section. If, for instance, a part of the body, such as the head, is exposed to a plane wave microwave field at 400 mc, the head may absorb 1.2 times the energy which is incident on its shadow cross section.

TABLE II

RELATIVE ABSORPTION CROSS SECTION OF A HOMOGENEOUS ISOTROPIC SPHERE COMPUTED FOR VARIOUS CIRCUMFERENCE-TO-WAVELENGTH RATIOS

Part of doll subjected to incident plane wave	Relative absorption cross section per cent				
	Natural size		Scaled to a height of 177.8 cm (70 inches)		
	Doll 1 height = 38.1 cm	Doll 2 height = 50.8 cm	Doll 1 height = 38.1 cm	Doll 2 height = 50.8 cm	Doll 3 height = 87.0 cm
Front or back	46	58	57	56	50
Side	56	48	55	59	50
Top	90	92	92	97	83
Applicable frequency mc/s	2880	2860	617	823	1409

This means that when objects are small compared with the wave-length, small changes in object size or aspect may cause large changes in the absorbed energy.

It appears that the curvature of an object does not substantially affect absorption characteristics, at least as long as the radius of curvature is larger than the wavelength. This statement applies particularly to mankind. Since the human body cross section is almost $1m^2$ (10,000 cm^2) for wavelength values smaller than about 60 cm (frequencies above 500 mc), high local possible absorption by structures, such as the nose or ears, should not contribute to the total absorption noticeably in view of the small local volume involved. This does not mean, however, that these parts may not be damaged seriously while the body is affected negligibly.

The experiments conducted by Anne, Salati, and Schwan resulted in the following conclusions:

(a) The relative absorption cross section of the human body is near 50 percent and appears to be independent of polarization of the incident field.

(b) Side and front aspects of the human body show essentially the same relative absorption cross section.

(c) The head, or top, aspect of the human body shows a relative absorption cross section of about 90 percent.

DIELECTRIC PROPERTIES OF BODY COMPONENTS

T. S. England and N. A. Sharples of the Royal Victoria Infirmary in England reported in 1949[22] on studies made of the dielectric properties of various components of the human body in the microwave region of the electromagnetic spectrum. The studies were made with the aid of a grant from the North of England Council of the British Empire Cancer Campaign which explains the references made to malignant tissue. The specimens used in the tests were obtained in most cases from surgical operations. They were specially selected for homogeneity and cut into sections suitable for insertion in a wave-guide cell. In this cell, the specimen was trapped between a tight-fitting plug of polystyrene one half wavelength in thickness and a reflecting metal plunger driven on a micrometer screw thread. The cell was surrounded by a bath of water thermostatically controlled at a temperature of 37° C to an accuracy of ±1/2° C.

From measurements on the standing wave pattern setup in the waveguide feeding the cell, two constants A and B, for the medium were determined by a method similar to that used by Roberts and Von Hippel.[23] The parameter A is the absorption coefficient and is expressed in nepers/cm. The parameter B is the phase constant expressed in radians/cm, and is a measure of the velocity of the wave in the medium. To the first approximation the dielectric constant of the medium can be taken as proportional to B^2.

Several representative body constituents were measured in this way at a wavelength of 3 cm, several thicknesses from various specimens being used where practicable. A listing of the results obtained is given in Table III.

TABLE III
DIELECTRIC PROPERTIES OF VARIOUS COMPONENTS OF THE HUMAN BODY

Specimen	A (nepers/cm)	B (radians/cm)	Remarks
A. Normal			
Whole blood	2.90	13.5	Uncoagulated by 1/500 mgm herapin
Skin 1. Breast	2.65	12.1	Free of areolar tissue
2. Leg	2.55	11.8	
Muscle	2.50	11.7	Stripped of fascial sheath
(Skeletal)	2.70	11.6	
Sigmoid colon	2.75	13.5	Sections included muscle coats, but no serous coat
	2.70	13.4	
Taenia coli	2.85	13.6	
Thyroid gland	2.70	13.7	Hyperplastic
	2.85	13.5	
	2.65	13.5	
Prostate gland	2.70	13.7	Hyperplastic
	2.50	13.7	
Fat 1. Breast	0.49	4.0	Free of areolar tissue
2. Leg	0.44	4.0	
Bone (femur)	0.56	5.4	Specimens obtained from a post-mortem examination. Bone machined accurately to fit waveguide
	0.56	5.4	
Bone marrow	0.79	5.1	

TABLE III – Concluded

Specimen	A (nepers/cm)	B (radians/cm)	Remarks
B. Malignant			
Carcinoma of breast (scirrhus)			Sections were taken from four different tumors and all were confirmed scirrhus carcinoma of breast by histological examination. The only difference was in the degree of fibrous stroma present in each tumor section.
(a)	2.70	13.1	
	2.75	12.5	
(b)	3.0	13.1	
	3.2	13.0	
(c)	2.75	12.5	
(d)	2.65	12.5	

The results quoted in the above table should have a measurement accuracy better than ±5 percent in A and +1 percent in B; but the difficulty of repeating biological specimens accurately should be appreciated. Measured in the same equipment, the corresponding constants for distilled water at 37° C were A = 2.70 and B = 16.0, which are in good agreement with those obtained by Collie, Hasted, and Ritson.[24]

England and Sharples arrived at the following conclusions as a result of their experiences:

(1) The absorption coefficients at 3.18 cm wavelength for a large number of body constituents are substantially the same as that of water at the same temperature. Fatty tissue and bone have a considerably lower absorption, and are relatively transparent to the radiation.

(2) The phase constants for a large number of body constituents fall into a group with a rather lower average value than that of water. Small, but significant, differences occur within the group, presumably attributable to the variation of water content. This is demonstrated by results taken on whole blood and its products after centrifuging (Table IV).

(3) At this wavelength no significant difference may be expected in the behavior towards the radiation of malignant and non-malignant tissue. It is to be expected that the effect of the polar water molecules will overshadow any changes which may result from biologically different structures.

Considering 3 cm wavelength radiation falling normally from outside on an area of the body, which can be treated for this purpose as a uniform and infinitely extended medium with average values of the constants A and B, calculation shows that approximately one-third of the incident power will be absorbed by the body and the remainder will be reflected. The high value of the absorption coefficient indicates that the majority of the absorbed power will be dissipated initially in the first few millimeters of skin. The resulting heat distribution will normally be modified by conduction and by the action of the vascular system.

TABLE IV
DIELECTRIC PROPERTIES OF HUMAN BLOOD

Specimen	Corpuscle concentration (x 10^6 per c. mm)	Radians/cm	Dielectric constant relative to air
Blood serum	---	15.2	57
Whole blood	4.9	13.5	45
Corpuscles in high concentration	17.1	12.7	40

Dielectric Properties of the Human Body for Wavelengths of 1.27 cm and 10.0 cm

In 1950, England reported[25] the results of additional experiments to determine the dielectric properties of various components of the human body for wavelengths of 1.27 cm and 10.0 cm. He concluded that the measurements at the two wavelengths are the same as those for the 3.18 wavelength which were reported in 1949[22] and again demonstrated that many of the body tissues behave very similarly to water in respect to their dielectric properties; that fat and bone, having much lower water content, are relatively transparent to the radiation; and that malignant and normal tissues do not differ significantly. One difference noted-between tissue behavior relative to water at 10 cm wavelength and at the shorter wavelengths was that the absorption constant A tended to have a value approximately double that of water, indicating a significant contribution of the ionic conductivity absorption. Assuming equal contributions to total absorption at this wavelength by the polar water molecules and by ionic conductivity, which is in agreement with the values obtained from measurements at longer wavelengths (summarized by Osborne and Holmquest[26], the ionic conductivity contributions to power absorption in such tissues at wavelengths of 3.18 cm and 1.27 cm would appear, by similar calculation, to be about 10 percent and 2 percent, respectively, and would not be immediately evident in the measured constants.

Chapter 5

IRRADIATION OF THE HEAD

For obvious reasons, there is little data available irradiation of the human head with radio-frequency energy. The Italian physician Cazzamalli performed some experiments with human brains exposed to the field of a relatively powerful radio transmitter in the 1920's[6] and claimed to have observed changes in brain wave patterns. He also claimed that he could produce hallucinations but only in highly suggestible individuals. His experiments were, unfortunately, generally inconclusive.

AUDITORY RESPONSE

A. H. Frey of the General Electric Advanced Electronics Center recently announced[27] the discovery that the human auditory system can respond to electromagnetic energy in at least a portion of the radio-frequency spectrum. The response is instantaneous and occurs at power levels as low as 0.4 milliwatt per cm^2 average power. Responses have been obtained from 200 to 3,000 mc. Subjects reported the perception of a buzzing or knocking sound. Deaf subjects appeared to perceive the sound as readily as those with normal hearing.

Frey reported the following eight points of experimental evidence in connection with this phenomena:

1. When the lower half of the head was covered, including the maxillary dental area, the RF sound could be perceived. When the top half of the head was covered, the RF sound ceased. Thus, fillings in the teeth were not implicated.

2. With the transmitter antenna enclosed in a radome and not visible to the subject, the antenna was rotated at various rates. Thus, the RF beam swept by the subject several times a minute. The subjects invariably perceived when they were swept by the RF beam.

3. Subjects were blindfolded and the beam was broken repeatedly in an irregular fashion by interposing a screen shield between the source and the subject. The subjects report of when the sound was "on" and "off" correlated perfectly with the unshielded and shielded conditions.

4. Subjects were placed in pairs in the RF beam. A screen shield was placed between the source head of one member of each pair.

The RF immediately ceased for the shielded member of the pair. The RF sound continued for the unshielded member.

5. Ear plugs rated to attenuate sound an average of 30 db were placed in the ears of subjects in the RF beam. The subjects reported a reduction in ambient noise level and an increase in the level of the RF sound, the latter probably being relative.

6. One of the deaf subjects had an average air conduction loss of 50 db. Bone conduction was fairly good; at worst he was down 25 db. His high-frequency response was relatively good. He could hear the RF sound with power densities approximating that needed for threshold perception in normals.

7. When a screen shield was placed so that it reflected RF energy which had passed the subject, the subject reported an increase in volume of the RF sound.

8. The subjects reported, when asked to localize the source of the RF that the apparent source was a short distance behind the head. No matter how they were rotated in the RF field, they localized the source in the same place; behind the head.

Frey concluded from his findings that it is difficult to accept that the perception of the RF is induced by acoustic energy external to the tympanic membrane.

PULSED ENERGY SLEEP

Although pulsed energy sleep utilizes frequencies far below the range used for radars, it is an effect produced on the human brain by pulsed electric energy of comparatively low power should therefore be of interest to those concerned with the effects of radar on the human body.

A comprehensive report on this technique has been given by A. S. Burhan of the Rand Development Corporation.[28] The technique was apparently developed by the Russians about 1948 and has been used mainly as a form of mental therapy on 500,000 patients. The technique is called, in Russian, "Electroson," in German, "Elektroschlaf," in French, "Sommeil electrique."

Pulsed energy sleep is an entirely different phenomenon from electroshock electronarcosis. It is based the concepts of "parabiosis" and "protective inhibition." It develops as a reaction to the passage of a very weak alternating current, with some DC components, through the brain tissue. The intensity of the current is not higher than 0.2 ma with a duration of 0.3 millisecond. The pulse rate is varied between 1 and 100 pps according to the age of the patient and the type of case under treatment. It has been proved to be free from any side reactions such as intoxication, electric accumulation, opisthotonus, or adverse effects on the respiratory or cardiac centers.

"Parabiosis" is a term chosen by N. E. Wedenskiy, professor of physiology, Leningrad, U.S.S.R. to describe a nerve condition characterized by an apparent absence of the nerve's fundamental characteristics (irritability and conductivity).

"Protective inhibition" is a term coined by the Russian, Pavlov, to describe a state of inhibition of the cortex of the brain brought about by external stimulants.

The first clinical model of the electronic equipment designed to produce-pulsed energy sleep was built in Russia in 1950. Since then, several changes have been made in the basic circuitry transistorized units and portable models have been produced.

The pulses generated in these units are applied to the patients with orbital and occipital electrodes. The orbital electrode represents the cathode and the occipital electrode represents the anode. The occipital electrode is made in a bifurcated form to fit on the mastoid processes.

After the electrodes have been placed, the current is turned on and its intensity increased until the threshold of the patient is reached. This corresponds to a level at which the feels a slight prickling sensation or a sensation of crawling ants under the eyelids or in the back of the orbital cavity. A heaviness gradually develops in the eyelids, a slight dizziness and drowsiness begins to be felt, thoughts disappear, and the patient falls into a steadily deepening physiological sleep. In this state, the patient will be in a quiet, relaxed position, usually lying on his side. Respiration remains even but becomes deeper and glower the pulse rate slows a few beats per minute. Upon discontinuation of the pulses, the patient awakens in a few minutes.

Pulsed energy sleep shows a very close relationship with physiological sleep except that the pulsed energy produces deeper changes in the body than those which could be obtained by a few additional hours of sleep brought about by indifferent rhythmic stimuli. Even after a short and shallow sleep during the procedure of pulsed energy application, the patients feel refreshed and in good spirits just as after a deep and long night sleep.

Pulsed energy sleep has been used by the Russian investigators as a form of therapy for neuroses, schizophrenia, the choreic form of rheumatic encephalitis, presenile psychosis, hypertension, duodenal ulcer, toxemia of early pregnancy, and serious forms of headaches. It has proved to be a safe and effective means of treatment. Usually from the very first treatment, patients become calm, lively, cooperative, and critical toward their disease; hallucinations and suicidal tendencies disappear and disturbed night sleep becomes normal.

IRRADIATION OF THE HEAD OF A MONKEY[29]

Experiments to determine the effects of close-range exposure of the brain of a monkey to high intensity radio waves were conducted by the National Institute of Health in March 1959 and the results of the experiments were reported by Dr. Pearce Bailey, Director of the National Institute of Neurological Diseases and Blindness, in budget hearings before the House of Representatives Appropriations Subcommittee.

In the experiments, the monkey was fastened to a chair in a sitting position inside a drum-shaped cage which served as a resonating cavity to greatly strengthen the electromagnetic energy within the cage. A radio antenna fitted to the top of the cage pointed toward the monkey's head, in line with his brain stem — the central vital portion of the brain. The antenna was excited by an AN/GRC-27 ultra-high frequency transmitter which operates in the 225 to 400 mc range and has a peak output of about 100 watts.

When the transmitter was turned on, the monkey was apparently unaffected for a few seconds, then it became drowsy. After a minute or so, the monkey became agitated, moving its head from side to side. In another minute, there appeared unmistakable signs of some impending disturbance in the vital centers of the brain. Finally, the monkey was thrown into a major convulsion a few seconds before death occurred.

Examination of the brains of ten monkeys which died in the experiments revealed no pathological cause of death. Another ten monkeys, whose exposure was cut short of death, showed symptoms which resembled those of Parkinson's disease in human beings. Most survivors recovered completely.

ADDITIONAL EXPERIMENTS WITH MONKEYS

Further experiments in which the heads of monkeys were irradiated with microwave energy were reported in 1959 by Bach, Baldwin, and Lewis.[30] The animals used in the experiments were young Macaca rhesus monkeys mostly weighing 7 to 10 pounds. The usual exposure period was from 2 to 10 minutes. The shortest exposure leading to death was 2 minutes and 55 seconds. The longest single exposure (in the horizontal head position) was about 3 hours without noticeable effect on the animal.

The ultra-high frequency energy was supplied by a Collins T17 AGR transmitter which is a 100-watt ground-to-air transmitter operating in the range from 225.0 to 399.9 mc.

Most of the exposures were to continuous wave radiation, although 100-percent modulation with a 500 and 1,000 cycle sine wave was also employed. A crude form of pulsing, by overmodulation, was also done in a limited number of exposures.

Behavior of Subjects During Exposure

The animals displayed arousal and drowsiness which were cyclic in nature. During the drowsy periods they were motionless, tending to keep the whole body in a fixed position. This pattern was usually seen within 60 seconds of initiation of the exposure. The animal then might stare with a wide fixed gaze. Then agitation, beginning with rapid side-to-side head movements, would occur. These movements often ceased abruptly and the animal would be quiet and unresponsive to touch, pain, light, and sometimes to sound stimuli. Alternating periods of arousal and drowsiness usually occurred. Three animals were deeply anesthetized with phenobarbital, being quite unresponsive to pinching of the Achilles tendon and to deep pin pricks. However, they could be made to move about in the chair within a minute after radiation began. By alternately switching the transmitter on and off, one of these animals was brought to the point of successive arousal and complete relaxation, in a 20-second cycle, reacting like a puppet on the end of a string. This particular effect was elicited most readily at 389 mc.

Eye Signs

With continued exposure, the animal would develop sagging upper eyelids of both eyes. Suddenly he would open his eyes and stare upward. The pupils were usually small and equal. Then the eyes would begin to move independently, and the pupils would dilate. Often one pupil would be larger than the other and in some instances lose its roundness. The pupils would then dilate and constrict irregularly and rapidly. Rapid, involuntary oscillation of the eyeballs would then occur, accompanied by rapid blinking. The involuntary oscillation of the eyeballs often persisted for several minutes after cessation of radiation.

Accompanying the eye signs were autonomic changes. The skin of the face would often become flushed and then pale. The nose often became pink and the respiratory rate increased. Salivation and lacrimation were also observed. With further exposure, the rapid blinking progressed to clonic movements of the eyes, bilateral clonic movements of the other facial muscles, a severe grimace which pulled back the lips from the teeth, clonic flexion of the neck, and symmetrical clonic movements of the upper extremities, trunk, and lower extremities in that order.

The onset of the rapid blinking and the grimace which heralded the generalized seizure was always a serious sign, although several animals with such signs progressed to complete recovery.

Motor Loss

Two animals developed paralysis of all four limbs. Two others developed weakness of the upper extremities and several developed an inability to coordinate voluntary muscular movements for varying periods. In all of these there also developed lesions of the occipital muscles overlying skin. One animal developed a right facial weakness with a concurrent anesthesia in the distribution of the upper two branches of the trigeminal nerve.

IRRADIATION OF THE HEADS OF DOGS

Searle, Imig, and Dahlen of the University of Iowa conducted experiments on dogs anesthetized with sodium pentobarbital and exposed to 2,450 mc - cw radiation for periods of one to seven hours.[31] A clinical, microwave diathermy machine (Raytheon Microtherm, Model CDM-IO with "C" director) was used to supply the microwave energy.

Healthy, mongrel dogs weighing 11 to 15 kg were anesthetized with 35 mg/kg of sodium pentobarbital and placed in a prone position on a light wooden frame. The head was supported by a gauze sling under the mandible. The rectangular director was placed with its long axis parallel to the midline of the calvarium. In this position the antenna was 5 cm from the surface of the scalp and a near field or, at most, crossover field, was considered to exist at the scalp. Power densities calculated on this are approximations. Depending upon the percentage of the maximum power output of the diathermy unit used, the calculated field intensities at the scalp varied from 0.5 to 0.8 watt/cm^2.

Copper-constantan thermocouples were used for temperature measurements which were read out on a multi-channel recorder. Intracranial temperatures were obtained by passing the thermocouples into the brain through small holes drilled in the skull. The cisterna magna temperature was obtained by passing a thermocouple through a hollow needle inserted between the first and second cervical vertebrae. Thus, simultaneous recording of temperature was obtained from the frontal lobe (thermocouple inserted through the frontal bone to one side of the midline), the midbrain (thermocouple inserted from the side through the temporal bone) and the cisterna magna in close association with the medulla and roof of the fourth ventricle. Rectal temperatures were similarly obtained at a point about 6 cm inside the external anal sphincter. Skin temperatures were measured on the irradiated scalp surface. Temperature measurements were made with the microwave generator turned off.

Pressure transducers with amplifiers and recorders were used to measure spinal fluid pressure from the cistertna and arterial pressure from the femoral artery. Heart and respiratory rates were obtained from the blood pressure or spinal fluid pressure tracings.

The power output of the generator was monitored or varied to prevent, as nearly as possible, skin damage. Experiments have shown that about 42° C is the threshold of skin temperature above which, with the type of radiation used, thermal damage to the skin will occur over periods of exposure of about 2 1/2 hours. Ten dogs were subjected to power densities averaging 0.5 watt/cm^2 at the scalp for 150 minutes. The purpose of the experiment was to determine the pattern of temperature changes at the three intracranial sites and at the relatively distant but physiologically associated, site in the rectum. The results are shown in Table V.

TABLE V

TIME-COURSE OF TEMPERATURES AT VARIOUS SITES IN TEN DOGS SUBJECTED TO IRRADIATION OF THE HEAD WITH MICROWAVE ENERGY MONITORED TO PREVENT EXCEEDING 42° C AT THE SCALP

Time	Temperature, °C				
Minutes from onset of exposure	Scalp	Frontal lobe	Mid-brain	Cisterna magna	Rectum
Control	34.1	38.4	38.5	38.2	38.1
10	40.0	39.1	39.3	40.1	38.7
20	40.1	39.6	39.6	40.2	39.1
30	40.3	39.9	40.0	40.3	39.4
40	40.9	40.2	40.3	40.7	39.7
50	41.0	40.3	40.4	40.8	40.0
60	41.0	40.6	40.4	40.9	40.1
70	41.1	40.6	40.6	41.3	40.3
80	41.4	40.7	40.7	41.1	40.4
90	41.5	40.8	40.7	41.3	40.5
100	41.6	40.9	40.9	41.3	40.6
110	42.2	40.9	41.0	41.4	40.7
120	42.0	40.9	40.9	41.4	40.9
130	41.9	41.0	41.0	41.4	40.9
140	42.3	41.0	41.1	41.5	41.0
150	42.5	41.0	41.1	41.6	41.1

As seen in Table V, the temperature was increased more rapidly in the cisterna magna than in any of the other sites where it was measured; with the exception of the skin. Although the temperatures measured in the brain, cisterna, and rectum were initially similar, the temperature in the cisterna was increased more rapidly and to a greater extent than in the other locations during the irradiation.

As might be expected, this experimental procedure had no serious effect upon the animals beyond the hyperthermia and minor skin damage to the scalp. The results suggest that the relatively slowly circulating fluid within the cisterna accumulated heat more rapidly than the other brain tissue sites which represented well-vascularized tissue.

With the thought that perhaps brain cells in relatively close approximation to the fluid-filled ventricles of the brain or to the subdural fluid-filled areas such as the cisterna magna might suffer damage at temperatures similar to or greater than those of the first experiment, a second more severe heating regimen was used. In this experiment 12 dogs were exposed to the maximum output of the generator for 1 hour; and six dogs were exposed to the same output for 2 hours. Chemical tests of the cerebrospinal fluid were made which indicated that not only did damage to brain cells probably not occur, but there was essentially no leakage from the blood into the cerebrospinal fluid.

A final series of tests was carried out in the same manner with two dogs except that attention was directed to femoral arterial blood pressure, spinal fluid pressure, respiratory and heart rates, and rectal temperatures. The experiments showed a rapid heart rate during the entire irradiation period which was thought to have resulted, in part at least, from the effect of the sodium pentobarbital anesthesia maintained throughout the period. The decrease in blood pressure which occurred near the end of the exposure could be attributed almost entirely to an effect resulting from the irradiation. The decrease in blood pressure which occurred unaccompanied by a decrease in heart rate suggested a decrease in cardiac output, a net vascular dilation, a reduced blood volume, or elements of all three factors. This shock-line trend occurred without any visible sign of central nervous motor excitation of a convulsive nature.

IRRADIATION OF THE HEADS OF RATS

Dr. M. L. Keplinger of the University of Miami has reported[32] on animal experimentation involving irradiation of the heads of rats with microwave energy at 2,400 mc. A pulsed magnetron was used having an average power output of about 20 watts,

Signs of Intoxication

Obvious signs of microwave effects were observed in rats exposed on the head at close range (3.8 cm from the antenna). The rat was immediately aware of some type of pain stimulus tried to move to avoid it. There was squealing and struggling within 15 to 25 seconds. The ears were hyperemic at first, then turned dark in color. First, second, and third degree burns were eventually produced on the skin directly in front of the antenna. The most conspicuous effect was stimulation of the central nervous system with muscle spasms, tremors, and chronic convulsions. The tail stood up almost straight. This stimulation was so marked that it aroused a rat from deep surgical pentobarbital anesthesia. Periods of central stimulation alternated with periods of depression. However, the periods of depression grew shorter as exposure time increased.

When the rat was moved farther away from the antenna (3 in. or 7.6 cm), there was a similar central nervous stimulation when the rat was exposed on the head, but when the lumbar region was exposed at this time distance, central stimulation did not then become apparent. It is also interesting that the approximate lethal exposure time for a rat exposed the head at a distance of 7.6 cm was 43 minutes; while it was 24 minutes for one exposed at the same distance in the lumbar region.

There was no cutaneous burning as the distance was increased to 10 cm or greater. At 12 cm or more, the obvious signs of microwave effects (tremors, etc.) were very slight, redness of ears and nose was still produced.

Gross Post-Mortem Findings

Exposure of the head in the near field (at 5.3 cm) caused hyperemia (congestion of blood) with petechial and some diffuse hemorrhages in the tissues under the skin. The skin was discolored (greenish-gray). Muscles of the head and neck (directly under the antenna) looked "cooked." The capillaries of the cerebral cortex and meninges were distended with blood and "leaking." In the lungs, there was marked congestion with hemorrhagic areas apparently caused by thrombic emboli. The heart was contracted and filled with blood clots,

Chapter 6

IRRADIATION OF THE TESTIS

EFFECTS OF MICROWAVE IRRADIATION AT 2880 MC (10.4 cm)

T. S. Ely and D. E. Goldman of the U. S. Navy Research Institute published in 1957[33] a preliminary evaluation of relative sensitivities of the whole body, the eye, and the testis in the human being using experimental data derived from experiments with animals. These relative sensitivities are listed in Table VI are based upon the thermal effects of microwave radiation only.

TABLE VI

RELATIVE SENSITIVITIES OF THE WHOLE BODY, THE EYE, AND THE TESTIS IN THE HUMAN

Structure	Initial temp. (°C)	Maximum temp. (°C)	Temperature rise (°C)	Steady state field (mw/cm²)	Cooling time constant (sec)
Whole body	37.0	39.0	2.0	100	50 joules/cm²)
Eye	37.0	45.0	8.0	155	100
Testis	35.6	37.0	1.4	5	250

Ely and Goldman irradiated the testes of dogs with microwave radiation at a frequency of 2880 mc (10.4 cm). The experiments were carried out on the dog because the species has testes comparable in size and to the human organ.

Steady state field intensities were obtained for the experiments and, in addition, an attempt was made to evaluate the effect of clothing and sweating on temperature. As was expected, clothing decreased the field intensity required to maintain 38° C and "sweating" (drops of warm water applied to the scrotum) increased it. When clothing and sweating were combined, the results were similar to those when neither was used.

Normal dog testicular temperatures ranged from 30.30° C to 36.25° C. Testicular temperatures in the exposed animals varied from 36° C to 44° C, and were maintained in all but one case for 60 minutes.

Most of the dog's testicles were removed surgically on the fourth day after exposure, the rest on the third and fifth day, and were studied microscopically. In all cases it was observed that the most severely involved areas were immediately adjacent to the capsule; the most central areas were either less severely or not at all damaged.

It was concluded from the experiments that testicular reactions to heat injury from a radar source are apparently basically the same as those due to hyperthermia associated with other conditions and many other causes. The minimal testicular damage is almost certainly reversible. Even considerably more severe testicular insult will probably be reversible, with the only finding being a temporary sterility. An even greater injury could result in permanent sterility.

It was also concluded that the testes would be the structures most likely to be damaged by microwave energy. They owe their special sensitivity to their physical location relative to the body surface, their poor ability to dissipate heat due to a poor or lacking vascularity, and high sensitivity to temperature increase. The undesirable effect of excess temperature on the testes is tubular injury.

In connection with the evaluation of the relative sensitivity of the testes, an effort was made to find the "normal" human testicular temperature, but it was found that testicular temperature varies considerably from time to time, depending on different environmental factors.

The determination of a temperature threshold for testicular damage proved to be a rather difficult problem. Ely and Goldman felt that it was undoubtedly a function of exposure time, and probably also of age. It also appeared normal body core temperatures are damaging to the testes of many species, although a well-defined time factor is not known for man.

EFFECTS OF MICROWAVE IRRADIATION AT 24,000 MC (1.25 CM)

Gunn, Gould, and Anderson of the University of Miami reported in 1960[34] on experiments in which the testes of rats were exposed to microwaves of 1.25 cm wavelength (24,000 mc) at a power density of 2500 mw/cm. In the experiments, the testes of the rats were exposed at a distance of 7.6 cm from the antenna in an environment maintained between 24° and 25° C. Exposures were for 5, 10, or 15 minutes. Following exposure, the animals were sacrificed and the testes were examined grossly and microscopically for damage.

Results of 15-Minute Exposures

Six days following 15-minute microwave exposure, there were severe third-degree burns of the scrotal skin. The testes showed many opaque areas, hemorrhage, and collapse. Microscopically there was extensive coagulation necrosis of the seminiferous tubules. Interstitial and vascular tissues were also involved in the necrosis.

Twenty-nine days following 15-minute exposures the scrotal skin was healing. The testes were fibrotic and reduced in size. Microscopically, the tubular outlines were devoid of germinal epithelium. The interstitial tissue contained numerous fibroblasts. Very few Leydig cells were present.

Results of 10-Minute Exposures

Six days following 10-minute microwave exposure, the scrotal skin showed small areas of third-degree burn. The testes were small with a few opaque areas. Microscopically, the testes showed focal areas of coagulation necrosis. The majority of tubules showed moderate to severe degeneration with only occasional normal tubules present. Of marked interest was the apparent normal appearance of the interstitial tissue.

Thirteen days following 10-minute microwave exposures, the scrotal skin burns were healed. The testes were small, with opaque areas present. Microscopically, the process of repair was in progress. Tubular debris was not present. Tubules were showing regeneration of cellular elements with active spermatogenesis present. Interstitial tissue showed hyperplasia with normal appearing cellular elements.

Twenty and twenty-nine days following 10-minute exposures, the histologic picture of the testes was essentially the same as thirteen days after exposure.

Results of 5-Minute Exposures

Six days following 5-minute microwave exposure, most animals showed no damage to the scrotal skin. A few animals showed small second-degree burns. The testes were enlarged with marked pallor. Microscopically, all testes moderate to severe edema. Most testes showed no tubular damage. A few testes had small areas of tubular degeneration. Interstitial tissue was apparently unaffected.

Thirteen days following 5-minute exposure, the scrotal skin was normal; and grossly, the testes appeared normal. Microscopically, the testes showed slight to moderate edema. The tubules and interstitial tissue were normal in appearance.

Twenty-nine days following 5-minute exposures, the testes were histologically normal.

COMPARISON OF EFFECTS OF MICROWAVE RADIATION AT 2450 MC AND THE EFFECTS OF INFRARED RADIATION

Experiments designed to compare the effects of microwave radiation and infrared radiation on the testes were conducted by G. W. Searle and his associates at the State University of Iowa and reported in 1960.[59] In these experiments, the testes of young, adult, male, albino rats were exposed to the output of a clinical microwave diathermy apparatus (Raytheon Microtherm, Model CDM-10 with "C" director). The rest of each animal was shielded by means of a copper screen. The exposure was made with the distal end of the testis directed toward the antenna at a distance of 5 cm. One testis was used for recording the temperature of the distal, middle, and proximal part of the organ approximately in the longitudinal axis by means of needle thermocouples. The other testis was used for follow-up studies of histological damage.

A similar arrangement was used with an infrared lamp (Burdick, "Zoalite" Z-12) as the source of radiation. In this case an asbestos shield was used to protect the rest of the animal from the radiation.

The stages of histological damage to the testes were classified as follows:

STAGE 1 - Diminished germinal elements in some seminiferous tubules, sloughing of germinal elements, and intertubular edema with enlargement of intertubular spaces.

STAGE 2 - Some seminiferous tubules devoid of germinal elements. Germinal elements may take on a "coagulated" appearance.

STAGE 3 - More tubules completely sterile.

STAGE 4 - Nearly complete degeneration of most of the seminiferous tubules with only Sertoli cells and basement membrane remaining. Presence of multinucleated "giant cells" with dark staining nuclei.

One of these stages of damage was assigned to each testis examined on the basis of microscopic examination of several fields in slides of stained longitudinal sections.

Each irradiation modality was varied in power output to maintain as nearly as possible, constant temperature for a 15-minute period. The central thermocouple was used as a guide in this. Thus, infrared and microwave exposures were equilibrated by the temperature effects.

In the first series of rats the stages of histological damage were assessed one hour after the irradiation period (15 minutes exposure at various intratesticular temperatures). Two to five rats were used in these experiments and the results were as follows:

    A. Intratesticular temperature 30° C, no damage from microwave or infrared radiation.

    B. Intratesticular temperature 35° C, Stage 1 Pius damage to three rats from microwave, damage from infrared.

    C. Intratesticular temperature 37° C, Stage 1 plus damage to four rats from microwave, no damage from infrared.

    D. Intratesticular temperature 38° C, Stage 2 plus damage to two rats from microwave, no damage from infrared.

A second series of rats was similarly exposed (15-minute exposure at various intratesticular temperatures) except that they were sacrificed and the testes examined two days after exposure rather than one hour after exposure as in the first experiment. The damage by microwave exposure was more extensive by histological criteria after two days than after one hour. The damage by infrared radiation after two days was much less than that resulting from microwave exposure. The specific results were as follows:

    A. Intratesticular temperature 35° C, Stage 2 minus damage to eight rats from microwave, Stage 1 minus damage to three rats from infrared.

    B. Intratesticular temperature 36° C, Stage 2 minus damage to seven rats from microwave, Stage 1 minus damage to two rats from infrared.

    C. Intratesticular temperature 37° C, Stage 2 plus damage to six rats from microwave, no damage from infrared.

    D. Intratesticular temperature 38° C, Stage 3 plus damage to seven rats from microwave, Stage 1 minus damage to two rats from infrared.

E. Intratesticular temperature 39° C, Stage 4 damage to six rats from microwave, Stage 1 minus damage to two rats from infrared,

COMPARISON OF THE EFFECTS OF MICROWAVE RADIATION AT 24,000 MC AND THE EFFECTS OF INFRARED RADIATION

Gunn, Gould, and Anderson of the University of Miami reported in 1960[34] on experiments designed to compare the effects of microwave radiation at 24,000 mc and infrared radiation on the testes.

The testes of rats were exposed to microwaves of 1.25 cm wavelength (24,000 mc) at a power density of 250 mw/cm^2. The testes were exposed at a distance of 7.6 cm from the antenna in an environment maintained between 24° and 25° C. One group of rats was exposed to microwave for five minutes, at the end of which time the intratesticular temperature developed was 41° C. Another group of rats were exposed to infrared for five minutes. The distance from the infrared lamp to the scrotal area was so adjusted that within five minutes of exposure an identical maximum intratesticular temperature of 41° C was developed.

Two weeks following exposure, the animals were sacrificed and the testes were examined grossly and microscopically. The histologic appearance of the testes of microwave and infrared exposed groups was identical. The only histologic alteration noted was a slight edema.

These experiments led Gunn, Gould, and Anderson to suspect a possible effect of microwave radiation on the male endocrine system. Their experiments in this area are discussed in the following chapter.

Chapter 7

EFFECTS OF MICROWAVE IRRADIATION OF THE TESTES ON THE ENDOCRINE SYSTEM

The work of Gunn, Gould, and Anderson of the University of Miami on the effect of microwave radiation on the male endocrine system of the rat reported in 1960[34] was stimulating and indicative of the type of experimentation that could be expected to open up broad, new areas of investigation.

Their primary purpose was to determine if exposure of the testes of rats to microwaves of 1.25 cm wavelength (24,000 mc) at a power density of 250 mw/cm² would produce a functional disturbance in the male endocrine system even though there might be no structural evidence of damage to the testes.

The endocrine glands involved in the study were the pituitary, the testes, and the dorsolateral prostate. In the course of studies on the rat prostate it had been shown that intracardiac administration of tracer doses of the radioisotope Zn-65 concentrates to a high degree in the dorsolateral lobes of the prostate, paralleling the rich natural zinc content of the gland.

It had also been learned from earlier experiments that the capacity of the dorsolateral prostate to concentrate Zn-65 is controlled by the hormones testosterone and gonadotrophin. A simplified diagram of the pituitary-testes-prostate endocrine chain illustrating the hormonal control of Zn-65 uptake by the dorsolateral prostate is given in Table VII.

A fall in the uptake of Zn-65 by the dorsolateral prostate indicates that either the pituitary, the testes, or the prostate have been damaged. To determine whether the dorsolateral prostate itself has been damaged, testosterone is administered. If the prostate does not respond with increased Zn-65 uptake, the gland has been damaged.

If the administration of gonadotrophin results in no increase in Zn-65 uptake by the dorsolateral prostate, definite damage to the testes is indicated.

If the administration of both gonadotrophin and testosterone is required to cause an increase in Zn-65 uptake by the dorsolateral prostate, damage to the pituitary gland is indicated.

In the microwave exposure experiments, the testes of the rats were exposed at a distance of 7.6 cm from the antenna in an environment maintained between 2° C and 25° C. Exposures were for 5, 10, or 15 minutes. At various times following exposure, the rats were injected with tracer doses of Zn-65, the animals were sacrificed, and the amount of Zn-65 taken up by the dorsolateral prostate was determined. At the same time, the testes were examined grossly and microscopically for damage.

TABLE VII

SIMPLIFIED DIAGRAM OF THE PITUITARY-TESTES-PROSTATE ENDOCRINE CHAIN

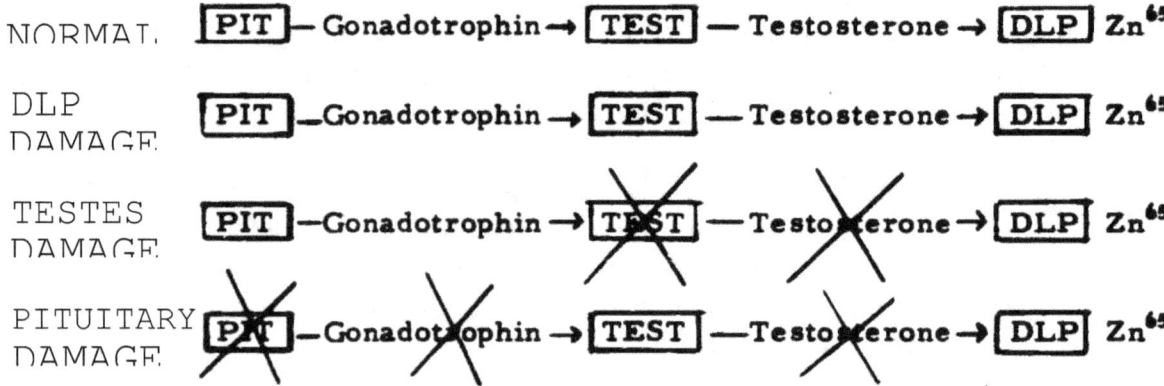

ABILITY OF DORSOLATERAL PROSTATE TO CONCENTRATE Zn-65 AFTER EXPOSURE

After 15-Minute Exposure

Six and 29 days following 15-minute exposure to microwaves, the capacity of the dorsolateral prostate to concentrate Zn-65 was 60 percent to 70 percent lower than control values. Glandular weights were 20 percent lower than controls.

After 10-Minute Exposure

Six days following 10-minute exposure to microwaves the capacity of dorsolateral prostate to concentrate Zn-65 was diminished 35 percent.

Thirteen, 20, and 29 days after exposure Zn-65 uptake was depressed 55 percent to 60 percent.

There was a lowered in dorsolateral prostate weight 6, 13, and 20 days following 10-minute microwave exposure.

Twenty-nine days after exposure, glandular weights were 29 percent lower than control values.

After 5-Minute Exposure

Six days following 5-minute exposure to microwaves, the capacity of dorsolateral prostate to take up administered Zn-65 was depressed 30 percent.

Thirteen days after exposure, Zn-65 uptake was diminished 45 percent.

Twenty-nine days after exposure, Zn-65 uptake did not differ significantly from the unexposed control value.

At no time following the 5-minute exposure was any significant alteration in the weight of the dorsolateral prostate. However, although there no evidence of structural damage after the 5-minute exposures, there was interference with the male endocrine system.

LOCATION OF DAMAGE IN THE ENDOCRINE CHAIN

After 10-Minute Exposure

In the 10-minute exposure groups, which showed Zn-65 uptake 50 percent below normal, the administration of testosterone restored Zn-65 uptake to control levels, whereas the administration of gonadotrophin was completely ineffective. Dorsolateral prostate weights, depressed in the microwave-exposed, untreated rats, were restored to normal limits by testosterone, but not by gonadotrophin. The experiment indicated twofold damage as follows:

a. Damage to the pituitary gland indicated by insufficient gonadotrophin output.

b. Damage to the testicular interstitial indicated by the failure of the tissue to fully respond to the hormone being elaborated.

After 5-Minute Exposure

In the 5-minute exposure studies, which showed Zn—65 de— pressed 45 percent from normal, administration gonadotrophin or testosterone restored the Zn-65 to control levels. The experiment indicated that both the prostate and interstitial tissue of the testes were able to function, and the cause of the lowered Zn-65 uptake was due to a lack of pituitary gonadotrophin indicating damage to the pituitary gland.

Dorsolateral prostate weights, although not depressed in the microwave-exposed, untreated rats, were significantly increased above normal values by the administration of either hormone.

A transient fall in Zn-65 uptake with a later return to normal levels, and the type of response to hormones noted in the 5-minute microwave exposure experiments, were very similar to effects noted with sublethal exposures to X-rays.

MICROWAVE EXPOSURE VERSUS INFRARED EXPOSURE

The next question investigated by Gunn, Gould, and Anderson was whether the structural changes noted in the testes, and the endocrine disturbances seen following exposures to microwaves, were really due only to thermal effects or due to a combined thermal plus some unknown effect inherent in the microwave range used.

In order to explore the question, simultaneous experiments were set up to compare the effects following microwave and infrared exposures. One group of rats was exposed to microwaves for five minutes, at the end of which time the intratesticular temperature developed was 41° C. Another group of rats was exposed to infrared for five minutes. The distance from the infrared lamp to the scrotal area was so adjusted that within five minutes of exposure an identical maximum intratesticular temperature of 41° C was developed.

Two weeks following exposure, both groups were injected with tracer doses of Zn-65, the Zn-65 in the dorsolateral prostate was determined, and the testes were examined grossly and microscopically. The histologic appearance of the testes of microwave and infrared exposed groups was identical. The only histologic alteration noted was a slight edema.

The results of the Zn-65 uptake studies showed a 45 percent fall in Zn-65 uptake in the microwave-exposed group, denoting a break in the pituitary-testes-prostate endocrine chain. In the infrared-exposed group, however, Zn-65 uptake was not altered from control levels, indicating that at this temperature range there was no damage to the male endocrine system.

These experiments indicated that there is a marked difference in the actions of microwaves of 24,000 mc and of infrared on the male endocrine system. The results are supported by work of Steinberger and Dixon[35] and Elfving[36] in the same field and suggest an athermal effect of microwaves.

Chapter 8

IRRADIATION OF THE EYE

The human eye is one of the few organs in the body which can be exposed to radiation directly rather than through intervening skin and varying amounts of adipose tissue. For this reason, and because of its particular structure, the eye is particularly susceptible to damage from microwave radiation.

The vitreous fluid contained in the eyeball reacts to heat in much the same manner as the white of an egg in that it becomes opaque and the process is irreversible. The crystalline lens of the eye has been shown to be peculiarly susceptible to the effects of radiated energy, whether ionizing, infrared, or radio frequency, all of which cause the development of opacities (cataracts) in this normally transparent component of the eye.

Considerable experimental work has been conducted in connection with the study of experimental radiation cataracts induced by microwave radiation. One of the primary investigators in this particular area has been Dr. R. L. Carpenter of the Department of Biology, Tufts University. Other investigators who have made important contributions include Richardson, Dune and Hines, Williams, Monahan, Nicholson and Aldrich, Daily, Wakim, Herrick, Parkhilt, and Benedict. The reports by Carpenter in 1958[37] and 1959[38] are particularly comprehensive.

The experiments of Carpenter were performed with rabbits. Rabbits were chosen as subjects because the diameter of the rabbit eye is three-fourths, and its volume approximately one-third that of the human eye. In addition, the body temperature of the rabbit, measured rectally, is 38.7° C (101.6° F), compared to the normal human rectal temperature of 37. 5° C (99.6° F).

Carpenter's microwave source was manufactured by Raytheon and was based on their Model CDM-4 Microtherm with additional circuitry to provide either continuous or pulsed wave at a frequency of 2450 mc. The output could be pulsed at duty cycles of 0.5 percent to 66 percent. The pulse could be varied between 50 microseconds and 2000 microseconds and the pulse repetition rate between 140 and 2200 per second. The antenna used was a Microtherm Director "C", a corner reflector type. The instrument was powered from a voltage stabilized line.

A directional couple, in the coaxial cable to the antenna permitted leading off a small, known fraction of the power output to a Hewlett Packard No. 430C microwave power meter and thus provided constant monitoring of power. An S-band silicon microwave diode on the back wall of the exposure chamber was connected to a Tektronix oscillograph for checking the pulse width, shape and amplitude.

All exposures were carried out in a chamber 36 by 34 by 20 inches which was lined with microwave absorbent material. The animal was placed in this anechoic chamber in a copper-lined wooden box with only its head exposed to the microwave field. The corneal surface of the eye was positioned exactly opposite the dipole crossover of the antenna and two inches distant from the surface of the plastic housing which covered the dipole.

Measurements of the power density of the microwave field were made calorimetrically in the anechoic chamber at the position of the rabbit eye, with the rabbit box in place. The maximum output of the microwave generator yielded a power density of 400 mw/cm^2 at the 2-inch position of the eye.

Changes in the temperature of the eye during irradiation were measured by means of hypodermic needle thermistor of 22 or 24 gage connected to a balanced bridge circuit which fed into a four channel, strip-chart recorder. For body temperatures, a thermistor enclosed in vinyl plastic at the end of a one-eighth inch, vinyl plastic covered lead was inserted in the animal's rectum. Accurate temperature records could thus be obtained for the entire period of irradiation, as well as before and after.

TIME AND POWER THRESHOLDS FOR INDUCTION OF LENS OPACITIES BY CONTINUOUS WAVE RADIATION AT 2450 MC

Carpenter performed 136 experiments to determine the minimum single exposure period which would cause an opacity to form in the lens. The results of these experiments form the basis for Figure 1, in which the duration of single exposures is plotted against the applied power density. The broken line on the curve represents a projection which may or may not be valid for power densities below 120 mw/cm^2.

CUMULATIVE EFFECTS OF REPEATED SUBTHREHIOLD EXPOSURE PERIODS

Carpenter reported in 1958[37] that at a power density of 289 mw/cm^2, lens opacities developed as the result of repeated exposures of relatively brief duration; any single one of which was not sufficient to cause an opacity. The minimum exposure employed was 3 minutes. The shortest single exposure period which has proved to be cataractogenic at this power density is 15 minutes.

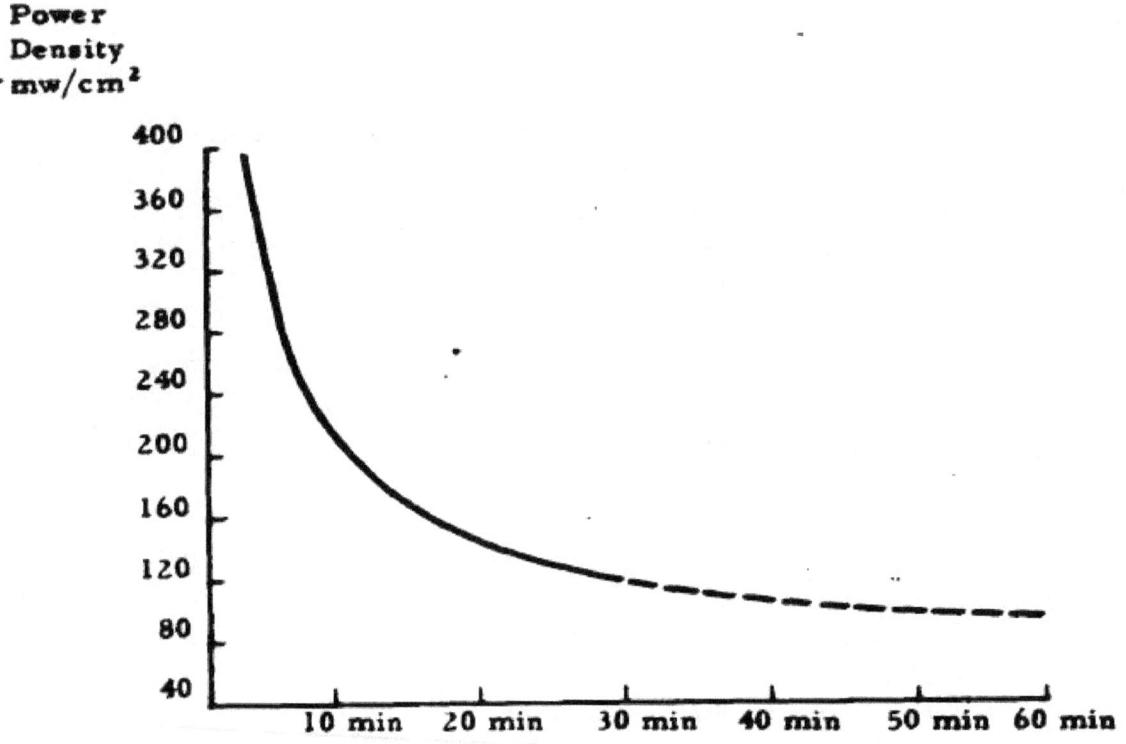

Figure 1. - Time and power thresholds for the production of lenticular opacities in rabbits exposed to 12.3 cm microwaves.

Carpenter reported in 1959,[38] that at a power density of 280 mw/cm², exposures of 3 minutes made daily for five successive days caused opacities in every me of five cases. The same results were obtained in five cases in which the eye was exposed to radiation three times for three minutes each time with the exposures being made every fourth day. When the interval between the three-minute exposure periods was increased to a week, however, the lens remained unaffected after five successive exposures in every one of the five experiments performed.

Carpenter felt that these results suggest if the cataractogenic factor of microwave radiation is one that initiates a chain of events in the lens, the visible and end result of which is an opacity, this factor must initially require an adequate power density acting for a sufficient duration of time in order to start the chain of events. If either the power density or the duration are less than a certain threshold value then the damage done to the lens is nor irreparable and recovery can take place provided that sufficient time elapses before a subsequent, similar insult. In these experiments, it appears that the interval necessary for recovery after the damage done by a three-minute exposure at 280 mw/cm^2 must be greater than 4 days, but need not be longer tun one week. Carpenter emphasised that this statement applies only to the conditions described, namely, a three -minute exposure at 280 mw/cm^2. It has previously been shown that if the exposure period is four minutes at this power density, then lens opacities may result from two exposures given a week apart or from two or three exposures separated by two-week intervals. It has not yet been determined what the requisite recovery period must be following a 4-minute exposure at this power density.

Carpenter referred to the "cataractogenic factor of microwave radiation" but was not red to state what this may be. It has been quite generally assumed that the effect of microwave radiation on living tissue is entirely a thermal one, an effect of the heat that results from absorption of RF energy by the tissue. It is certainly true that localized heating occurs as a result of absorption of microwave energy, yet one is reminded that thermal effects usually occur at or above a specific temperature, such as the melting point, boiling point, flash point, etc., and not as the result of intermittent excursions to temperatures below critical value. It is possible to conceive of a thermal effect which could be the result of either a high temperature for a short time or a lower temperature for a longer time; but it is difficult to imagine the same effect resulting from a low temperature occurring for a short time but repeated at widely separated intervals.

At a power density of 280 mw/cm^2, the shortest single exposure period which will cause a lens opacity is 5 minutes, at which time the temperature of the vitreous body at the posterior pole of the lens has reached 49.3° C. At the end of a 3-minute exposure period, however, this temperature is only 47.2° C. Inasmuch as this duration of exposure, if repeated at four-day intervals, causes a lens to form, but if repeated at weekly intervals has no effect, then one can support a "thermal effect" viewpoint only by assuming that a lesser temperature increase may be cataractogenic if it occurs frequently enough. Such an assumption would be difficult to support, for if the temperature is further reduced and the frequency of its occurrence increased, the constant normal body temperature will eventually be reached which is certainly not a cause of cataract.

A cumulative <u>thermal</u> effect as the cause of microwave induced opacities becomes even less probable in the light of further experiments. Recognizing the valid criticism that 280 mw/cm² is still, from the biological viewpoint, a fairly high power density, and that a vitreous temperature of 47.2° C, even though occurring ever so briefly, is, nevertheless, well outside the range of a rabbit's normal or even its pathological variation, experiments were undertaken in which the eye would be repeatedly exposed to radiation at relatively low power densities.

Having ascertained that at a power density of 120 mw/cm² the minimum was 35 minutes, eyes were exposed at this power density for periods of 30 minutes repeated at two-day intervals. Of the four experiments completed, lens opacities developed after two such exposures in one case and after three exposures in three cases. In a rabbit under sodium Nembutal anesthesia, the temperature of the vitreous body at the end of 30 minutes of irradiation at this power density was 44° C.

A point worth mentioning with regard to these experiments was that at 120 mw/cm², the animals remained quiet in the microwave field without anesthesia. In the four experiments, two animals were irradiated with anesthesia. In these two cases, therefore, we can be sure that the heat-dissipating vascular system was not functionally affected by an anesthetic and that the ocular temperature was surely not higher than 44° C and was probably less than that. The animals did not appear to be experiencing any discomfort whatever.

Carpenter continued his series of experiments to include daily exposures of the eye to one hour of continuous wave radiation with the power density reduced to 80 mw/cm² and in some cases to 40 mw/cm² without anesthesia.

Of the three experiments attempted at the 80 mw/cm² level, two animals died accidentally of causes not ascribable to the radiation. The third animal was irradiated for an hour each day for five consecutive days each week, this schedule being maintained until a total of 19 hours of irradiation had been given. At that time, slit-lamp and ophthalmoscopic examinations were negative; the lens was clear and without sign of opacity. Examined 13 days later, however, the irradiated eye showed a well developed cataract in the posterior cortex. The non-irradiated eye was normal.

It is of interest to note that in the anesthetized animal, one hour of exposure to continuous wave radiation at 80 mw/cm^2 raises the temperature of the vitreous body to 42.8° C, which is only 4.1° C above body temperature. Carpenter felt that if further experiments at the 80 mw/cm^2 level also result in cataracts, it will have to be concluded that if they are induced as a thermal effect of microwave radiation, then it is a thermal effect requiring neither a critical temperature nor even a marked elevation of temperature in the tissue.

EFFECTS OF PULSED MICROWAVE RADIATION

With pulsed radiation, the eye can be subjected to rapidly repeated peaks of high microwave power while the average power during the exposure period remains relatively low. Because the thermal flux is identified with the average power alone, Carpenter hoped, through a series of experiments, to discover whether microwave energy might exert other than a thermal effect.

He reported in 1958[37] on the first 16 of his initial group of 25 experiments then in progress. In these experiments, the eye was exposed to pulsed microwave radiation at an average power density of 140 mw/cm^2 with pulse peaks of either 280 or 560 mw/cm^2. The duty cycle employed was therefore 50 percent or 25 percent; the latter being the lowest duty cycle he could obtain with the equipment then being used. In 62 percent of the experiments, lens resulted from exposure periods and associated ocular temperatures significantly less than those required for induction of opacities by continuous wave radiation of identical average power. He therefore suggested that the cataractogenic effect of microwave radiation might not be primarily a thermal one, and advised giving attention to peak powers of the microwave field when assessing the possibility of hazards to personnel.

Carpenter felt that further experiments along this tine would be fruitful if he could employ pulsed radiation having a greater spread between peak power and average power. This would demand much lower duty cycles than he was then able to attain. He therefore had his equipment redesigned to allow duty cycles as low as 0.5 percent. He reported in 1959[2] that with this equipment he irradiated 15 animals using average power densities ranging from 120 mw/cm^2 down to 40 mv/cm^2

and with accompanying peak powers of 400 mw/cm² up to 800 mw/cm². Under these conditions, opacities developed in 53 percent of the experiments.

For example, opacities occurred after 60-minute or longer exposures of the eye to pulsed microwave energy when the average power density was only 80 mw/cm² but the peak power was 400 mw/cm². A 45-minute exposure had no effect. He considered it to be significant that at the end of a 1-hour exposure period at the 80 mw/cm² average power density, the temperature within the eye had risen to 42.8° C, only 4 degrees above body temperature. He also noted that this same power density has no effect when applied as continuous wave radiation for a 1-hour period. He felt that these results pointed to peak power as being an important factor in causing opacities to develop when the eye is exposed to pulsed microwave radiation.

NON-COMPLEMENTARY EFFECTS OF MICROWAVE RADIATION AND X-RAY

Lens opacities induced by microwave radiation and those caused by ionizing radiation are similar in several respects. Typically, both develop in the posterior subcapsular cortex; changes occur initially in the region of the posterior suture; and frequently the opacity takes the form of striate masses concentric with the equator, which Later migrate axialward to form ring-shaped cataracts. One marked difference is in respect to their latent periods. Cogan and Donaldson have shown[39] that after single doses of 1200 to 1500 r. of X-ray at 1500 KV, opacities develop in the lens after 25 to 30 days. In contrast, Carpenter found that following exposure to microwave radiation, opacities appeared after latent periods of 1 to 8 days, the average time being 3-1/2 days.

In order to test whether these two types of radiation could complement each other in their cataractogenic effects, the eyes of 22 rabbits were exposed to X-ray of 200 KV. Both eyes received equal doses of 1500 r. Either 24 hours or one week later, the right eye was exposed to a single, subthreshold dose of microwave radiation (280 or 352 mw/cm² for 3-1/2 minutes). In all of these experiments, opacities appeared in both eyes at the same time and were of similar degree. Under the conditions of the experiment, therefore, microwave radiation did not act in any complementary or supplementary manner to shorten the latent period for formation of X-ray induced opacities, nor did it affect the extent of the damage to the lens by x-irradiation.

COMPARISON OF OPACITIES RESULTING FROM MICROWAVE AND INFRARED RADIATION

If the opacities formed in the posterior cortex of the lens are solely a thermogenic effect of microwave radiation, and therefore are purely a result of hyperpyrexia, then it should be expected that any influence causing a comparable rise in ocular temperature would produce a comparable result. Hartman has stated[40]: "The physiological and pathological reactions to hyperpyrexia alone, induced by whatever mechanisms, ouch as hot baths, heated cabinets, diathermy, heat stroke, thermal burns, and microwaves are essentially comparable. Imig[41] plotted temperature gradients by 12.25 cm. microwaves in excised beef eyes and found that the point of maximum temperature was in the vitreous body at the posterior surface of the lens, and hence agreed with the site of damage.

Carpenter undertook to study the effect of increasing the ocular temperature by means other than microwave radiation. The beam from a 4.5 amp. carbon arc source, passed through an infrared filter transmitting only wavelength from 1000 to 3000 millimicra, was focused either on the anterior or on the posterior capsule of the lens in the anesthetized animal.

With focus on the anterior capsule and a 6-minute exposure, the cornea was severely burned with subsequent ulceration, scarring, and vascularization. Although the lens was thereby almost obscured, it remained possible to identify an opacity on or beneath the anterior capsule. In most of the experiments, focus was on the posterior capsule, so that the cornea was spared from burn and suffered only a transient clouding. With duration of exposure of either 5 or 10 minutes, opacities resulted in every case <u>but they were always located in the anterior cortex of the lens, despite the infrared being concentrated at the posterior capsule</u>.

The results are in agreement with various other observations on cataracts caused by infrared radiation. The opacities occur typically in the anterior cortex and in this respect differ from those caused by either microwave or ionizing radiation.

STUDY OF POSSIBLE INTERFACE EFFECT AT LENS-VITREOUS BODY BOUNDARY

Several suggestions have been offered to why opacities induced by microwave energy occur typically in the posterior cortex of the lens, just beneath the capsule. One suggestion was that there might be an interface effect at the lens cortex-posterior capsule boundary or at the capsule-vitreous body boundary, with resulting reflection of power to cause a concentration in the posterior cortex. If so, the temperature might well be higher in the lens cortex than in the vitreous body just behind the lens, which is where ocular temperature measurements are usually made. Carpenter therefore performed the following experiment.

In an anesthetized animal, a suture was placed around the tendon of the inferior rectus muscle at its insertion on the eyeball. Traction on the suture rolled the eye upward to an extent such that the cornea disappeared behind the superior orbital margin and the inferior surface of the sclera was presented between the open eyelids. With this surface positioned two inches from the antenna, the eye was irradiated for 12 minutes at 280 mw/cm^2 power density. The orientation of the lens with respect to the microwave source was now reversed; the power had to pass through the sclera, vitreous body and posterior capsule before reaching the lens cortex - an approach from the rear. If an interface effect should in fact exist, Carpenter reasoned that the changed orientation might so alter it as to affect the results.

Opacities appeared as usual in the posterior subcapsular cortex in all three surviving cases. Carpenter concluded that if there is an interface effect, it is not a critical factor with respect to the site of the opacity.

EFFECT OF DISTANCE OF EYE FROM MICROWAVE SOURCE

It was also suggested that at a 12.3 cm wavelength, the 2-inch distance at which the eye was usually positioned for irradiation might be an important factor; perhaps introducing the effect of a half wave length distance between the antenna and the lens and so promoting creation of standing waves. The distance from the dipole to the cornea was 6 cm. and from the cornea to the posterior lens surface 8mm making a total of 6.8 cm.

To test whether this particular distance was significant, Carpenter irradiated eyes in 11 rabbits for 15 minutes at 250 mw/cm^2 power density. In six cases, the distance to the eye was increased from two to three inches but the power density at the position of the eye was kept constant at 240 mw/cm^2. In all 11 cases, the same type of opacity appeared in the posterior lens cortex. A similar series of experiments at power density of 220 mw/cm² gave further evidence that the two-inch distance was not of itself significant in the induction of opacities by 12.3 cm microwaves.

CHANCES IN THE ASCORBIC ACID CONTENT IN LENSES OF RABBIT EYES EXPOSED TO MICROWAVE RADIATION

The work of Carpenter prompted the later work of L. O. Merola and J. H. Kinoshita of the Harvard Medical School in collaboration with Dr. Carpenter to determine the biochemical changes in lenses of rabbit eyes exposed to microwave radiation. The results of these experiments were reported in 1960[42].

Rabbit eyes were irradiated with microwave radiation at a frequency of 2450 mc, a wavelength of 12.3 cm, and a power density of 280 mw/cm² in the manner previously used by Dr. Carpenter. The length of exposure was varied depending on the severity of opacification desired.

The effect of microwaves on the permeability of the lens was first studied. Previous investigations had established that changes in permeability of the lens alter the sodium and potassium content. The lens has a typical intracellular composition characterized by high potassium and low sodium content. An increase in permeability would be expected to produce a drop in K and an increase in Na as the lens equilibrates with ions of the intraocular fluids. For this reason, the lenses of eyes exposed to microwave radiation were examined for changes in K and Na levels. Since only one of the rabbit eyes was exposed to radiation, the lens of the other eye served as a control.

The experiments showed that levels of K and Na in an irradiated lens were only altered when an obvious opacification had occurred. Where the opacities were minimal, the cation content was essentially normal. The more severe the opacification, the greater was the shift in the cation distribution of the lens. A marked illustration of this was shown in the mature cataracts where the cation balance was grossly altered. In instances where transparency was maintained after irradiation, the levels of K and Na were normal. These results indicated that the factors which influence the cation distribution of the lens were not particularly sensitive to microwave irradiation.

A number of other constituents of the lens were also equally unaffected. The levels of protein thiol groups, ammonia, glucose, and lactate fell only after an opacity had developed. These results suggested that the first sign of damage to a lens by microwave radiation was the formation of opacities rather than any change in chemical constituents.

This view was upheld until the levels of glutathione and ascorbic acid were examined. Glutathione occurs in the lens in a higher concentration than in any other body tissue. Ascorbic acid is another reducing substance also found in high levels in this ocular tissue. The reason these two substances are found in such significant quantities and the role they play in the physiology of the lens are facts as yet understood.

Ascorbic acid and glutathione in the lens appeared to be more sensitive to microwave radiation than any of the other chemical constituents studied. However, the results of a large number of experiments conclusively showed that the drop in ascorbic acid occurred before any change in glutathione. Furthermore, the fall in ascorbic acid was observed before the development of opacification. The eyes of rabbits were exposed for 8 minutes and removed 18 hours after irradiation. The lenses of the microwave-exposed eyes were transparent, had a normal content of glutathione, but had a substantially lowered level of ascorbic acid. The results seemed to indicate that ascorbic acid is the most sensitive chemical constituent to be affected by microwave irradiation--of the lens. This conclusion is noteworthy because in other forms of experimentally produced cataracts, such as in ionizing irradiation and diabetic cataracts, a decrease in glutathione content has been shown to be the first observable change. The distinguishing feature of microwave cataracts, then, is the fall in ascorbic acid content as the first sign of damage.

During the exposure period the possibility existed that an increase in temperature was the cause of the observed drop in ascorbic acid. This compound, being fairly unstable, could conceivably be affected by the increase in temperature incurred upon exposure of the lens to microwave energy. Experiments were therefore designed to check this possibility. Removal of the lenses one-half hour after the microwave exposure revealed that no change in the ascorbic acid content had occurred. This finding ruled out the possibility that an increase in temperature during exposure was the cause of the disappearance of ascorbic acid. The drop in the level of ascorbic acid was not observed 6 hours after irradiation, but was found in lenses removed 18 hours after exposure. This indicated that the decrease in ascorbic acid which results from microwave irradiation of the lens does not occur immediately after exposure but that it develops after a latent period of six to eighteen hours.

Chapter 9

THE EFFECT OF MICROWAVES ON UNICELLULAR ORGANISMS

The consensus of opinion throughout the years since microwave energy was first introduced has been that no effect is produced other than that due to the production of heat. To a great extent, investigators have tended to ignore some early observations of possible non-thermal effects. This chapter reviews some of these observations before examining more recent experiments.

REVIEW OF PAST EXPERIENCES

The first observation of possible non-thermal effects of exposure to microwave energy was made by Muth in Germany in 1927[44] when he described the formation of chains of emulsified fat particles in electromagnetic fields. Krasny-Ergen analyzed these observations in 1936[15] and explained the phenomena by the formation of electrical dipoles on each particle followed by mutual electrostatic action..

Another interesting report was made in 1946 by Nyrop of Copenhagen, Denmark[9], who described four years of systematic investigation of the effect of high frequency electric currents on biological objects with a view to finding a specific effect apart from the heat effect generally supposed to be the only effect of high frequency currents on tissue, bacteria, virus, etc.

Nyrop experimented with current at a frequency of 20,000 kc. By means of a modulator, the current was periodically turned off and on with a frequency of 10 - 100 kc in such a manner that the recurring current pulses were separated by current-free periods; each being n times longer than the single current impulse. With the modulator employed, n could be varied between 3 and 20.

Nyrop reasoned that heat developed by such a modulated current had a greater possibility of dissipation than heat produced by an unmodulated current. With a certain cooling of the object it was thus possible to adjust n so that the temperature of the object, through which the current passed, was kept below the temperature at which heating effects might occur. Liquids could be passed intermittently between electrodes and cooled between each passage but, as a too rapid flow would make it difficult to secure a uniform velocity of all parts of the liquid and thereby a uniform treatment, it was also important, in this case, to modulate the electric current. With a modulated current, the actual time of treatment to be dealt with would be only the time when current was passing through the object; namely, l/nth of the total time of treatment.

Nyrop exposed Bacterium coli in a liquid medium to the modulated current and reported that 99.5 percent of the bacteria were killed in 7 seconds when the field strength was 230 volts/cm. At 288 volts/cm, the time required was only 4 seconds. He reported no marked difference whether the treatment took place between 12° C and 40° C or between 40° C and 60° C. Using an improved apparatus, he reported 99.6 percent of the bacteria killed in 5 seconds at 205 volts/ cm and 99.98 percent in 10 seconds using the same field strength. A similar effect produced by heat would have required 600 seconds at 60° C.

Foot-and-mouth disease virus was completely inactivated when exposed to 260 volts/cm for 10 seconds with the temperature not above 36° C. The virus was completely inactivated in 2.4 seconds when exposed to a field strength of 480 volts/ cm. To inactivate the same virus by heat required 60 hours at a temperature of 37° C.

A virus inactivated by heat can be used as a vaccine. Nyrop discovered that a virus inactivated electrically showed no vaccinating effect and therefore reasoned that the electrical treatment acted on the virus molecule differently from heat treatment.

In experiments with tissue cultures, Nyrop demonstrated that it was possible to kill the tissue in 300 seconds when using the modulated current with a field strength of 22 volts/cm without raising the temperature of the tissue above 30' C.

Of interest in the discussion of the effects of microwave energy on cell structures is the observation of W. J. V. Osterhout made in 1949[3]: "It seems desirable to stress again the importance of the non-aqueous surface films of living cells. These structures, invisible under the highest magnification, play an all-important role. They are the seat of considerable electrical forces so that when a sufficient number of ceils is in series, an e.m.f. of 500 volts may be available, as in the electric eel. The high electrical capacity and resistivity of the cell is due to them. They regulate the intake and outgo of all substances and thus control metabolism. If they are destroyed death ensues at once. The behavior of such films involves the little understood laws of surface chemistry and physics and deserves intensive study."

Also of interest is the statement of K. S, Cole made in 1949[45]. "It is well established not only that the interior of the living cell is very different from the external inanimate environment in composition, structure and electric potential, but also that these differences are maintained by a barrier at the surface which is necessary for the life of the cell.... The cell membrane is thought of as a 'leaky condenser'. All living cells have a membrane capacity of the order of one microfarad per square centimeter. The membrane has a thickness of about 30 angstroms."

EXPERIMENTAL CELL RESEARCH USING PULSED ELECTRO, MAGNETIC FIELDS

Additional investigation has been made recently into possible nonthermal effects of electromagnetic fields by Teixeira-Pinto, Nejelski, Cutler, and Heller of the New England Institute for Medical Research at Ridgefield, Conn. Their work, published in 1960[46], was essentially a continuation of the investigations of Muth, Krasny-Ergen[15], Liebesny[47], and Schwan[48] and the later studies of Wildervanck, Wakim, Herrick, and Krusen[49] using pulsed electromagnetic fields in place of continuous fields.

From the standpoint of biological investigations, the pulsed field is preferable to a continuous field since the production of heat is minimized. Investigations into the possible significance of the alignment of particles into chains, as a non-thermal effect of high energy, high frequency fields in biological situations were reported in several preliminary papers from this New England laboratory.[50, 51, 52]   One of the phenomena noted was that motile bacteria were constrained in their motion in such electromagnetic fields. This observation led to the investigation of the response of various higher unicellular organisms. As a result, a series of interesting phenomena were observed.

Description of the Experiments

The experiments utilized a breadboard assembly of electronic components constructed in the laboratory. It consisted of a Hartley circuit oscillating at from approximately 0.1 to 100 megacycles per second, suitably modified to deliver maximum field gradients of several thousand volts, peak-to-peak, per centimeter across the enclosure serving as the load. The oscillator was 100 percent modulated by means of a General Radio Pulser No, 1217-A with several stages of amplification between the pulser and the oscillator. With lower field gradients, an unmodulated carrier was also employed for studies principally involving non-viable material. Voltages were measured by a General Radio Vacuum Tube Voltmeter, No. 1800-B. The waveform was monitored by a Tektronix oscilloscope No, 545.

The energy from the radio frequency source was coupled either directly or through a link from the tank circuit of the oscillator and brought into two electrodes which could be placed in a variety of configurations. One of the most commonly used assemblies was similar to one suggested by Herrick[53] and was constructed by painting electrodes on a microscope slide with silver, printed-circuit-paint, leaving an air gap of from 2 to 5 mm between the electrodes. The preparations to be examined were placed between two coverslips separated by a gasket of silicone grease and the sandwich was placed upon the electrodes. Mapping of the electromagnetic lines of force indicated that a rather homogeneous field existed between the electrodes in the horizontal plane.

An insulating plastic stage which could be placed on top of the metal stage of the microscope was made, with a hole for optimal light transmission. There was no contact between the electrodes and the media or material exposed to the field. When larger enclosures were required to expose the root tips of plants or Drosophila, lucite chambers were constructed with metallic electrodes in their walls covered by 0.1 mm glass cemented to the electrodes. The glass-covered electrodes were separated by an air gap of from 4 mm to 1-1/2 cm.

Iron filings, starch particles, colloidal carbon particles of sizes ranging from 0.5 micron to 400 micra, homogenized milk, oil suspensions in water, and water-suspensions in oil were studied at a variety of frequencies and field intensities. The substances studied were placed in various media; including air (in Drosophila experiments), water, oil, methyl cellulose solution, glycerol, acetone, sucrose, and dextrose solutions, as well as various protozoan and bacterial media. A particularly useful substance for study was a series of excellent suspensions of colloidal polystyrene spheres where the particle size was quite precise with a very small standard deviation. Polystrene spheres of 0.5, 0.8, and 1.171 micra were used. Most colloids were in an aqueous suspension, but other materials were substituted for the suspending phase where feasible. The biological substances used included mammalian erythrocytes, macrophages, E. coli, Cl. botulinum,
E. typhi, B. sutills, the root tips of common garlic (Allium sativum), Drosophila melanogaster (wild type), various species of Paramecia,
Amoeba proteus, Amoeba limax, Euglena gracilis, and other Euglenoidina, Actinophrys sol, Volvox, Pandorina, Eudorina, and Chlamydomonas.

Results of the Experiments

One of the first phenomena observed when particulates were placed in the electromagnetic field was the orientation of asymmetric particles with their long axes along the lines of force, followed by chain formation. Symmetric spheres can show no preferential orientation since there is no long axis; hence, only chain formation could be observed when using symmetric spheres.

Early observations indicated that the orientation and chain formation phenomena were frequency dependent as well as voltage dependent. Thus, when a suspension of polystyrene was exposed, and the effect of a variety of frequencies explored, it was found that there was one frequency band where optimum chain formation occurred with a minimum of voltage between the electrodes. As the frequency was raised or lowered from this optimum frequency band, the amount of voltage required for chain formation, as judged by chain length and the time required for their formation, increased by a considerable factor. In several cases these frequency optima were found to be within a range where the dielectric constant for both particles and suspending media do not vary. For example, there is no significant change in the dielectric constants of polystyrene and water between 1 and 100 megacycles; and optimum frequencies for chain formation were found within this range.

As biological materials were studied, the frequency and voltage dependency became even more obvious. It had been previously pointed out that motile microorganisms, when placed in the field, were constrained to travel along the electromagnetic lines of force instead of in random directions. These lines were portrayed by taking microphotographs of the alignment of inert particles in the field. The lines also occurred around a single electrode with the other electrode at a virtually infinite distance.

As various frequencies were explored, it was found that most motile organisms traveled "east-west" (along the lines of force) at frequencies in the lower megacycle range for as long as the field was maintained. As the frequency was increased, however, the organisms pivoted 90^0 and moved "north-south" (across the lines of force). Different organisms took an east-west or north-south orientation at different frequencies. Thus, in a preparation of mixed Colpidium, Rhabdomonas incurva, and Astasia Klebsi at a frequency of 8.5 megacycles, an inter-electrode peak-to-peak voltage of 309 volts per centimeter was required to have all of the organisms traveling in an east-west direction. At 11.5 megacycles, it was necessary to use a voltage .of 1016 per centimeter to have Rhabdomorns traveling east-west and Astasis north –south, while Colpidium was almost random. At 27 megacycles, only 582 volts per centimeter were required to have all of the organisms traveling in a north-south direction. Hence, it was possible to use a frequency with a sample of mixed organisms where one type was going east-west and another type simultaneously going north-south.

Efforts to define specific east-west and north-south frequencies for different organisms were unsuccessful. For example, some Euglenae were placed between two microscope coverslips sealed with silicone grease. At 5 megacycles the Euglenae all went east-west, and at somewhat under 6 megacycles they made a 90° pivot and went north-south. Twelve hours later, the organisms still responded with an east west orientation at 5 megacycles, but the frequency had to be raised to 18 megacycles to make the cells pivot and orient north-south. The only possible explanation for this response was the fact that the metabolites, either in the cells or media, had been produced and destroyed in the intervening hours. Obviously, this was an extremely small change which caused major differences in terms of frequency response. This type of varying optima for orientation was seen in virtually all living cells, depending upon differences in the age of the preparation or in the composition of the media. The difference in frequency response was probably due to a change in the dielectric constant of ions in the solution as a function of frequency.[54] In view of such sensitivity to very minute amounts of ions, the orientation frequencies could be determined only for a specific sample.

In a given situation, in addition to east-west and north-south frequencies, "dead" and "confusion" frequencies could be observed. The "dead" frequencies were so named because of the lack of a visible response of a certain species at the same time, and in the same field, where other species were responding optimally. The "confusion" frequency range was so named because the organisms were obviously sensing the field; were disturbed, and sometimes spun around their centers like a pinwheel. In one case, at 100 megacycles, paramecia were seen to spin with tremendous velocity about their long axes while stopped or migrating in the north-south direction.

Effects on Intracellular Structure

An early observation in an immobilized paramecium showed that certain asymmetric cytoplasmic inclusions were oriented when the electromagnetic field was impressed and reverted at once to their original position when the field was released. This led to the assumption that it might be possible to affect selectively, as a function of frequency and voltage, certain structures within the cell. These initial observations on paramecia were subsequently verified with Amoeba proteus.

When an unattached amoeba, with pseudopods extended so that it had a long axis, was placed in a 5. megacycle field, it oriented east-west while asymmetric cytoplasmic inclusions oriented in the same direction. However, when the frequency was rapidly changed to about 27 megacycles, the amoeba immediately pivoted 90° to point north-south; but the cytoplasmic inclusions still oriented east-west within the organism.

The above phenomena in amoebae were obtained with moderate voltages of approximately 300 to 50 volts peak-to-peak across the electrodes with a 15-microsecond pulse duration and a pulse repetition rate of 500 to 1,000 per second. As the voltage was increased still higher, Amoeba proteus, but not Amoeba Limax, responded by retracting pseudopodic extensions becoming spherical. There was a pronounced flattening of the sphere with a tendency to form an ellipsoid with a major axis parallel to the field. The moment the field was removed, the spherical form returned. When the field was sufficiently powerful, the sphere ruptured, with extrusion of the cytoplasmic contents, followed by rapid orientation and chain formation of these contents. The observation that intracellular material could be manipulated suggested that genetic material might be affected by such fields. Garlic root tips in water and Drosophila in air were therefore exposed to various fields at different frequencies and voltages. The initial findings in this area of investigation were reported in 1958.[51]

When an electromagnetic field is distorted by an object having greater or less conductivity than the media, the field distortion extends for some distance on either side of the object. An example of this could be seen when observing the behavior of both amoebae and Euglenae together in a field. As the Euglenae came into a position relatively close to the amoebae, they were very rapidly drawn toward the amoebae and, once one had come virtually into contact with the wall of an amoeba, it rotated with an enormously rapid spin. A similar phenomenon was observed when the amoeba in the field occasionally lost the tips of two of its pseudopods on opposite sides of the organism. The fragments became spherical, migrated down the length of the amoeba, and immediately began to spin very rapidly. A drop in voltage could stop the spinning completely, and a reversion to the previous conditions could induce it again.

One other phenomenon was noted with Amoeba proteus. If the organism was exposed to considerable voltage after having assumed a spherical shape and then the field was turned off, the amoeba seemed no longer capable of normal, purposeful, pseudopodic motion. Instead, there were wavering cytoplasmic extensions together with an increase in the width of the hyalin layer.

In contrast to the above phenomenon observed with <u>Amoeba proteus</u> and Euglenae, the ability to cause microorganisms to orient either with or across the lines of force was a phenomenon which was reproducible without any apparent visible damage to the organisms; provided they were not subjected to any significant heating. Preparations were placed in the field intermittently for as long as four to five days with no evidence of damage.

The north-south orientation of living microorganisms was a phenomenon which was unique with the limits of the experimental materials and frequencies used. All other non-biological materials always aligned along the lines of force (east-west). When Euglenae were killed thermally by a few long pulses at a carrier frequency at which the protozoa moved north-south when alive, the organisms maintained their original orientation. The individual cells, however, no longer remained separated when dead but instead formed chains extending in an east-west direction with the long axis of each cell perpendicular to the chain.

If additional heat was introduced while still maintaining the same frequency, the organisms pivoted and lined up end-to-end in a chain which also was parallel with the field. If, on the other hand, the initial frequency was such that the Euglenae were going east-west and they were thermally killed, they immediately took up the end-to-end chain position. A north-south chain formation was never observed.

In contrast to the normal lining up to form chains of cells in mutual contact, which was seen in almost all unicellular, non-motile organisms, Chlorellae seemed to be a notable exception. Chlorellae formed a general east-west type of chain; although at low frequencies it looked more like a network with the predominant axis being east-west. In no case, however, was it possible to induce a Chlorella to come within several diameters of its nearest neighbor.

Chapter 10

GENERATION AND DETECTION OF IONZING RADIATION PRODUCED BY MICROWAVE EQUIPMENT

It was reported by J. L. Spencer, Cot G. u. of the U, S. Air Force in 1957" that there is a tendency among those concerned with the operation of electromagnetic radiating equipment to consider only radio frequency energy as dangerous, completely ignoring the equally dangerous ionizing radiation generated by the equipment under their control. The development of high-power radio frequency equipment is inevitably accompanied by large increases in emission of scattered ionizing energy, mostly in the form of X-rays and gamma rays. With radio frequency power tubes such as magnetrons, klystrons, and thyratrons there seems to be an almost linear relationship between applied plate voltage and the production of spurious X-rays. There is also a rather direct relationship between the physical characteristics of the power tube and its ability to emit ionizing rays. Those tubes which produce longer radio frequency wave lengths are potentially more hazardous for X-rays.

Investigations of the nature of X-ray production by hydrogen thyratrons, for example, has revealed some alarming data. The X-ray emission from a thyratron operated with a constant plate voltage of 30 kv and a pulse repetition frequency of 100 per second proved to be less than 20 mr per hour. A change in the pulse repetition frequency to 250 per second (other conditions remaining constant), resulted in a production of 260 mr per hour. A further increase in the pulse repetition frequency to 500 per second caused a rise in X-radiation to about 4,800 mr per hour. A significant factor in the tests is that the tubes were operating at considerably less than their capabilities.

Another source of potentially harmful ionizing radiation is the multitude of radioactive electron tubes that have come into common use. The U. S. Air Force inventories over 500 types of these tubes containing up to 10 microcuries of radioactive materials per tube. These materials include carbon 14, cesium 137, cobalt 60, nickel 63, and radium 226. Adequate disposal procedures for these tubes present several recognized problems; however of more probable concern to personnel are the compounded high levels of radiation that exist in storage areas. Constant area and personnel dosimetry must be provided for at all levels.

There are, basically two types of injuries that may be caused by ionizing radiation. The first is the somatic lesion or the injury caused to the organism or system exposed to the radiation which is not transmissable to subsequent generations. It has long been known that ultraviolet, and ionizing radiation, such as X-rays or gamma rays, have a carcinogenic or cancer-producing action which may be produced by a single intense exposure or by chronic exposure. The source of the radiation may be external or internal in the form of a radioactive isotope.

The second type of injury is the much more subtle genetic injury. Man is becoming increasingly aware of the disastrous effects overexposure to ionizing radiation may have on subsequent generations. By genetic injury is meant an alteration in inherited characteristics. These alterations or aberrations are usually referred to as mutations and they may be benign or malignant. Mutations represent a chemical change in the structure of the gene, the basic unit of inheritance. The exact nature of the change is as yet unknown. Mutations may occur spontaneously in normal individuals or may be induced by ionizing radiation or chemical agents such as mustard gas.

The manner in which the maximum permissible dose of ionizing radiation has been reduced over the past 60 years is an indication of the serious effects upon mankind that this type of radiation can inflict. According to R. S. Stone,[58] in 1902 Rollins suggested 10 r per day; in 1925 Lewis suggested 0.2 r per day; in 1936 the U. S. Advisory Committee recommended 0.1 r per day; in 1950, ICRP recommended 0.3 per week, and by 1957 the NCRP specified 5 rems per year.

In considering ionizing radiation as produced by microwave generators, any increase in temperature induced by the radio frequency energy must be considered. Increased temperature may play an important background role in the degree of susceptibility to ionizing radiation. Even before it was discovered that X-rays caused mutations, it was shown that a temperature increase of 10° C more than doubles mutation frequency.

A very excellent treatise on the generation and detection of pulsed X-rays from microwave sources was given by Mr. A. P. De Minco of the Rome Air Development Center at the 1960 Tri-Service Conference on the Biological Effects of Microwave Radiation and was published in the Proceedings of the Conference.[56] De Minco stated that many microwave generators now contain components whose X-radiation outputs far exceed any industrial X-ray generators and he discussed the generation and detection of the unwanted by-product as follows:

GENERATION OF X-RADIATION

High-power electronic tubes, ouch as klystrons, magnetrons, traveling-wave tubes, and high-voltage hydrogen thyratrons, possess the basic physical parameters which allow them to act as powerful X-radiation generators; namely, a beam of electrons traveling high speed toward an anode, or target, which is at a very high voltage, and the subsequent stopping of these rapidly moving electrons.

Devices employing these phenomena, and operating with an applied voltage of more than 15 kv, can be considered potential hazards. In high-power microwave generators, conditions and components exist wherein the production of soft X-radiation at levels as low as 15 kv, through the "intermediate" range and on up to "hard" X-radiation at 300 kv, is possible.

The Hydrogen Thyratron

The hydrogen thyratron consists of an indirectly heated cylindrical cathode with an inner and outer cathode shield and cathode baffle. The nickel grid structure consists of a cylindrical portion and a perforated grid disc with a solid grid baffle below it. The molybdenum anode is enclosed by the grid disc. The tube ordinarily is operated in a line-type pulse modulator circuit with resonant charging of the pulse-forming network to a high peak forward voltage.

The operating parameters of the 1257 hydrogen thyratron tube are: 38 kv peak voltage, 2000 amp peak current with a 2.5 microsecond current pulse at a repetition rate of 200 pulses per second. While the effective energy of this tube is rather low, 25 kev, a one amp average current would give rise to rather intense "soft" X-radiation. The major portion of the X-ray beam emanates in a circle through the screen mesh of the grid-anode region. There may be considerable variation in X-radiation from tube to tube under similar operating conditions due to variations in grid emission. It has been reported that the 1257 tube has emitted X-radiation intensities up to 10,000 mr per hour at a distance of one foot from the tube. On an average, however, 1200 to 1500 mr per hour would be a more common radiation intensity.

For the most aggravated conditions, 1/16-inch steel paneling would attenuate this radiation down to a fraction of 1 mr per hour. Leaded glass 1/4 inch in thickness would attenuate 10,000 mr per hour to a negligible quantity.

The Traveling-Wave Tube[57]

The traveling-wave tube consists of three basic sections, the electron gun, the slow-wave structure, and the beam collector. The electron gun consists of a heater, a cathode, one or more control elements-called grids, and an anode or accelerating electrode. The electrons emitted from the cathode are made to converge through the center hole of the anode into the region where interaction with the radio frequency wave occurs. The beam is confined to flow longitudinally by an external solenoid or periodically magnets. The low level RF is applied to the slow-wave structure by means of a connector near the cathode end of the tube. The effect of the interaction is to bunch the beam by retarding some electrons and accelerating others, depending on the relative phase of the RF electric field and the position of the particles. In order for gain to occur, there must result a net deceleration of the beam; that is, part of the kinetic energy of the electrons is extracted as an increase in amplitude of the radio wave. The spent electrons are then intercepted by the beam collector to complete the circuit.

Even in well designed tubes, some of the electrons fail to reach the collector. Those that strike the anode give rise to X-rays at the forward end of the tube. The fraction that is intercepted by the tube body also produces radiation, but this is generally absorbed in the structure itself. Sixty to ninety percent of the beam terminates at the collector, and it is in this region that most intense radiation can occur. Lead shielding is often required in this area when the beam potential exceeds 50 kv.

Studies of traveling-wave tubes operating at 30 kv showed that X-radiation was emitted in a narrow band radially from the anode at a radiation level of 1 to 10 mr per hour, and that with proper precautions traveling-wave tubes need not pose a serious hazard to personnel. Shielding with leaded glass or sheet metal generally provides sufficient protection.

The Klystron

The high-power klystron operates on the principle of velocity modulation. It consists of an electron gun, an RF section made up of a series of resonant cavities with drift tubes interposed, followed by an electron collector. The electron beam generated by the electron gun is focused through the RF section, usually by means of a magnetic field. The electrons are then dissipated in the electron collector.

Essentially, the klystron possesses the same basic physical conditions existing in an ordinary X-ray tube; namely, a beam of electrons traveling at high speed toward an anode, or target (in this case the collector), which is at a very high voltage and the stopping of these rapidly moving electrons.

One of the most serious mistakes, and the most common one, made by personnel operating or servicing microwave equipment is to assume that a klystron is not generating X-radiation if RF is not applied to the tube. The klystron will generate approximately the same X-radiation intensity without RF applied as it will with the RF drive in operation, although there is some evidence that the RF voltage added to the DC beam voltage will intensify the X-radiation, and that the RF beam spreads the beam of electrons as it travels down the body of the tube, so giving rise to many X-radiation beams as they strike random targets. <u>A KLYSTRON OPERATING WITH THE RF DRIVE DISCONNECTED IS A POWERFUL AND DANGEROUS GENERATOR OF X-RADIATION AND MUST NEVER BE APPROACHED IF IT LACKS ADEQUATE SHIELDNG</u>.

The determination of safe radiation protection (shielding) for a klystron can only be accomplished by actual experimentation. Conventional tables cannot be used because they predict X-ray intensity on the basis of monochromatic radiation. The klystron produces a continuous X-ray spectrum, and most of the radiation is less penetrating than those that correspond to the limiting 400 kv energy. An oscillating klystron also has groups of electrons accelerated to voltages approximately twice the beam voltage producing X-radiation of greater intensity and greater penetrating power. In addition, the velocity and space distribution of electrons in an oscillating klystron cannot be predicted with any degree of accuracy in the region beyond the third cavity.

In a tunable klystron, X-radiation output may be increased or decreased by going through the frequency range. The physical Location or the X-radiation can be changed by beam focusing and "peaking" with variable parameters.

DETECTION OF X-RADIATION

The accurate detection and recording of ionizing radiation in an admixture with radio frequency energies involves problems of dosimetry generally not encountered prior to the advent of high powered radio frequency generators. Tests performed in which commercial area and personnel dosimeters were subjected to an RF field demonstrated the marked influence this energy has on the ionization recorded. Relatively low power, as little as 1 mw/cm^2, caused the degradation or altering of the ionization measuring capabilities of the commonly employed types of personnel dosimeters and area survey meters.

If the detector and circuits were shielded from the RF energy, the shielding attenuated the lower energy X-radiation before it reached the detector, thus making the detecting instrument almost insensitive to the lower energy X-rays.

Gas filled X-ray detectors such as geiger tubes and ionization chambers are not reliable detectors of X-radiation from RF tubes because intense radiation generates enough ion pairs in the gas to seriously alter the electric field within the detectors, changing their sensitivity and possibly making them inoperative.

The film type dosimeter accurately records high levels of ionization even in very strong RF fields. Absolute dosage determinations over large ranges of radiation flux are possible with photographic emulsions. Large photographic films placed in the environment of high-power generators not only can indicate levels of radiation intensity but describe patterns as well.

Photographic film for personnel monitoring ionizing radiation produced by RF tubes, however, is not satisfactory. Most X-ray emitting electronic tubes generally produce small, well collimated beams emitted through small faults or openings in the tube body or shielding, and the probability of a narrow, intense beam of X-radiation from an RF tube striking a personnel film badge is very remote.

An encouraging development in the search for an adequate radiation survey instrument is the Radiacmeter ME-118 developed by the Rome Air Development Center and scheduled to become a standard item in the U. S. Air Force Inventory. This meter is a scintillation type detector, and its theory of operation is as follows: X-radiation impinges on a crystal (phosphor); the crystal fluoresces, producing light in the visible spectrum; this light emission is detected by a photo cell whose electrical serves as an input to electronic circuitry, where the signal is amplified and then fed into a device for visual indication.

BIBLIOGRAPHY

1. Editorial. Nature, vol 138, Oct. 3, 1936.

2. Report of Conference on Comparative Effects of Radiation, Univ. of Puerto Rico, Feb. 15-19, 1960. John Wiley & Sons, N. Y.

3. Osterhout, W.J.V.: "Some Bioelectrical Problems," Proc. National Academy of Sciences, 1949, vol. 35, pp. 548-549.

4. Curtis, W. E., Dickens, F., and Evans, S. F.: "The 'Specific Action' of Ultra-Short Wireless Waves," Nature, vol. 139, 1936.

5. Alm, H., Einführung in die Mikrowellen-Therapie. (Fürst, Berlin-Schönberg, 1958)

6. Cazzamalli, F.: Neurologica, vol. 6, 1925, p. 193; Quaderni Psichiat, vol. 16, 1929, p. 81; Neurolgica, vol. 16, 1935, p. 47.

7. Lakhowski, G.: The Secret of Life. Heinemann, London, 1939.

8. Everdingen, W. A. G.: Ned. Tijdschr. Geneesk., vol. 82, 1938, p. 1873; vol. 84, 1940, p. 4370; vol. 65, 1941, p. 3094; vol. 87, 1943, p. 406.

9. Nyrop, J. E.: Nature, vol. 157, 1946, p. 51.

10. Livshits, N. N.: Biofizika, vol. 3, 1958, p. 426.

11. Jaski, T., and Süsskind, C.: Science, vol. 133, 1961, p. 3451.

12. Krusen, F. H.: IRE Trans. on Med. Electronics, 1956, PGME-4.

13. Mumford, W. W.: Proc. IRE, 49, Feb. 1961, pp. 427-447.

14. Handelsman, M.: Proc. Tri-Service Conf. on Biol. Hazards of Mic. Rad., July 1957, pp. 23-31.

15. Krasny-Ergen, W.: Hochfrequenztechnik und Elektroaskustik, vol. 48, no. 4, Oct. 1936, pp. 126-133.

16. Vogelman, J. H.: Proc. Tri-Service Conf. on Biol. Hazards of Mic. Rad. Plenum Press, N. Y., July 1960, pp. 23-31.

17. Overman, H. S.: Proc. Tri-Service Conf. on Biol. Hazards of Mic. Rad. Plenum Press, N. Y., July 1960, pp. 47-54.

BIBLIOGRAPHY - Continued

18. Jaski, T.: Electronic World, June 1961, pp. 31-36.

19. Schwan, H. P.: Proc. Tri-Service Conf. on Biol. Hazards of Mic. Rad., August 1959, pp. 94-106.

20. Meahl, H. R.: Proc. 4th Int. Conf. on Med. Elec., July 1961, p. 229.

21. Anne, A., Salati, O. M., and Schwan, H. P.: Proc. 4th Int. Conf. on Med. Elec., July 1961, p. 225.

22. England, T. S., and Sharples, N. A.: Nature, vol. 163, Mar. 26, 1949, pp. 487-488.

23. Roberts, S., and Von Hippel, A.: J. Appl. Phys., vol. 17, 1946, p. 610.

24. Collie, C. H., Hasted, J. B., and Ritson, D. M.: Proc. Phys. Soc., vol. 60, 1948, p. 145.

25. England, T. S.: Nature, vol. 166, Sept. 16, 1950, pp. 480-481.

26. Osborne and Holmquest: "Technic of Electrotherapy," 1944, p. 446.

27. Frey, A. H.: Proc. 4th Int. Conf. on Med Elec., July 1961, p. 158.

28. Burhan, A. S.: Proc. 3rd Ann. Tri-Service Conf. on Biol. Eff. of Mic. Rad., August 1959, pp. 124-135.

29. Aviation Week, May 4, 1959, pp. 29-30.

30. Bach, S. A., Baldwin, M., and Lewis, S.: Proc. 3rd Ann. Tri-Service Conf. on Biol; Eff. of Mic. Rad., August 1959, pp. 8292.

31. Searle, G. W., Imig, C. J., and Dahlen, R. W.: Proc 3rd Ann. Tri-Service Conf. on Biol. Eff. of Mic. Rad., August 1959, pp. 54-61.

32. Keplinger, M. L.: Proc. 2nd Ann. Tri-Service Conf. on Biol. Eff. of Mic. Rad., Sept. 1958, pp. 215-239.

BIBLIOGRAPHY-Continued

33. Ely, T. S., Goldman, D. E.: Proc. Tri-Service Symp. on Biol. Eff. of Mic. Rad., July 1957, pp. 64-74.

34. Gunn, S. A., Gould, T. C., and Anderson, W. A. D.,: Proc. Tri-Service Conf. on Biol. Hazards of Mic. Rad., July 1960, Plenum Press, N. Y., pp. 99-113.

35. Steinberger, E., and Dixon, W. J.: "Some Observations on the Effect of Heat on the Testicular Germinal Epithelium", Fertil. and Steril., vol. 10, 1959, p. 578.

36. Elfving, G.: "Effects of the Local Application of Heat on the Physiology of the Testis," T. A. Sahalan Kirjapaino Oy., Helsinki, 1950.

37. Carpenter, R. L.: Proc. of 2nd Tri-Service Conf. on Biol. Effects of Mic. Energy, July 1958, pp. 146-168.

38. Carpenter, R. L.: Proc. of 3rd Tri-Service Conf. on Biol. Effects of Mic. Energy, Aug. 1959, pp. 279-290.

39. Cogan, D. G., and Donaldson, D. D.: "Experimental Radiation Cataract 1, Cararacts in Rabbits Following Single X-ray Exposures," Arch. Ophth, vol. 45, pp. 508-522.

40. Hartman, F. W.: "The Pathology of Hyperpyrexia". Proc. of 2nd Tri-Service Conf. on Biol. Eff. of Mic. Rad., July 1958, pp. 54-64.

41. Imig, C. J.: "Review of Work Conducted at State University of Iowa," Proc. of 2nd Tri-Service Conf. on Biol. Eff. of Mic. Rad., 1958, pp. 242-253.

42. Merola, L. O. and Kinoshita, J. H.: Proc. of 1960 Conf. on Biol. Eff. of Mic. Rad. Plenum Press, N. Y., 1960, pp. 285-291.

43. Merola, L. O., Kern, H. L. and Kinoshita, J. H.: "The Effect of Calcium on the Cations of Calf Lens," Arch. Ophthalmol., vol. 63, 1960, p. 830.

44. Muth, E.: Kolloid Z. Vol 41, 1927, p. 97.

BIBLIOGRAPHY-Concluded

45. Cole, K. S.: "Some Physical Aspects of Bioelectric Phenomena," Nat. Academy of Sciences, vol. 35, 1949, p. 559.

46. Teixeira-Pinto, A. A., Nejelski, L. L., Cutler, J. L., and Heller, J. H.: "Experimental Cell Research 20, 1960, pp. 548-564.

47. Liebesny, P.: Arch. Phys. Therapy, vol. 19, 1939, p. 736.

48. Schwan, H. P., in Licht, Sidney: Therapeutic Heat. Elizabeth Licht, Pub. New Haven, Conn., 1958, p. 55.

49. Wildervanck, A., Wakim, K. G., Herrick, J. F., and Krusen, F. H.: Arch. Phys. Med. vol. 40, 1959, p. 45.

50. Heller, J. H.: Proc. 3rd Int. Symp. on the Reticuloendothelial System, Ronald Press, New York.

51. Heller, J. H., and Teixeira-Pinto, A. A.: Research Bull. vol. 4, no. 10, 1958.

52. Nature, vol. 183, 1959, p. 905.

53. Herrick, J. F.: Personal Communication to J. H. Heller.

54. Williams, D. R., Monahan, J. P., Nicholson, W. J., and Aldrich, J. J.: IRE Transactions on Medical Electronics, vol. 17, 1956, PGME-4.

55. Spencer, J. L., and Knauf, Col G. M.: Proc. of Tri-Service Conf. on Biol. Hazards of Mic. Rad., July 1957, pp. 52-59.

56. De Minco, A. P.: Proc. of Tri-Service Conf. on Biol. Hazards of Mic. Rad., Plenum Press, New York, 1960, pp. 33-46.

57. Sparks, R. A.: Proc. 4th Internat. Conf. on Med Electronics, vol. 230, 1961, pp. 30-36.

58. Stone, R. S.: Radiology, vol. 58, 1952, p. 639.

59. Searle, Dahlen, Imig, et al.: Proc. Tri-Service Conf. on Biol. Hazards of Mic. Rad. Plenum Press, New York, July 1960, pp. 192-195.

www.ingramcontent.com/pod-product-compliance
Lightning Source LLC
Chambersburg PA
CBHW062333220526
45469CB00008B/2701